GRASSLANDS

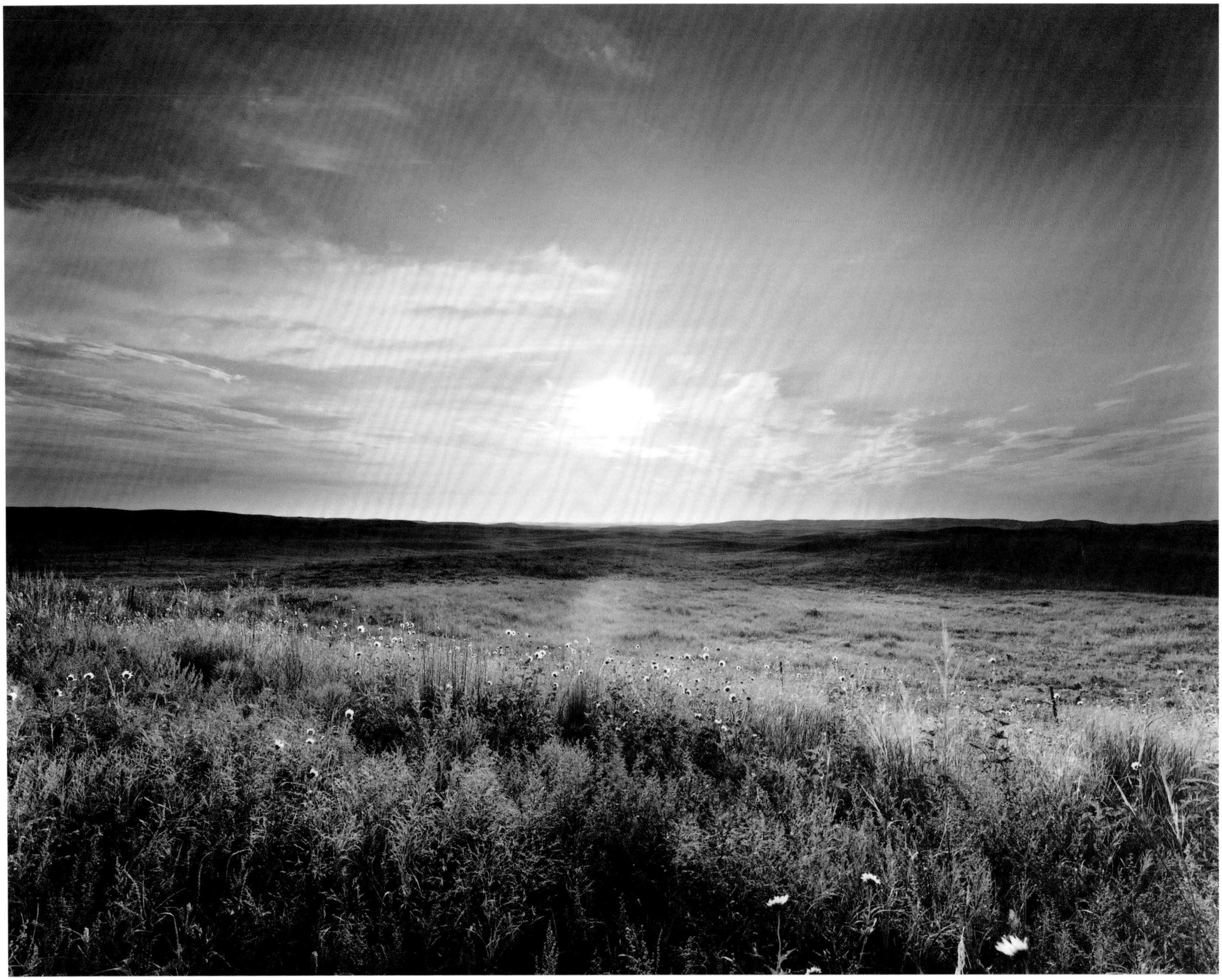

Don Kirby

GRASSLANDS

NAZRAELI PRESS

A timely vision

It is difficult to imagine scenes more politically relevant at this very moment than these photos from America's great grasslands. The pivotal period for most of these places occurred in a time that is a close parallel to our own, so grasslands stand as living, breathing records of an age of upheaval, economic depression and well-meant policy gone wrong. That they persist is testimony to a unique American strength: an evolved and evolving commitment to conservation. Yet they are at the same time not so much a recollection of the past, as they are an exhortation of what we must do to honor this legacy.

The idea of preserved grasslands in North America is traceable to 19th century figures such as the artist George Catlin, who wrote stunningly of creation of "a nation's Park" to protect the wonder of the Great Plains. Yet through accident, I happen to know the firm idea for achieving this aggregation spun from a single and singularly quirky brain of Henry Lantz, a man so obscure that I can find no one who knows what became of him. He was the extension agent in Phillips County, Montana and in the 1920s and '30s found himself surrounded by badly broken and busted dryland wheat farmers who, acting on their government's confident advice, had committed the sin of plowing what shouldn't have been plowed. The wages of these sins were meager and harsh for farmers and land alike. For the former, it was simple destitution and literal starvation throughout the Great Plains. For the land, it was the Dust Bowl, airborne real estate that landed in New York City, darkened the skies sufficiently to turn on streetlights in Washington, D.C. and coated the decks of ships at sea in the Atlantic.

Lantz knew for a fact the root problem was land too arid to tolerate the plow, friable, thin and parched. These places, the Great Plains, the Palouse, Great Basin, Colorado Plateau, these demand grass and punish severely anyone who acts contrary to this demand. Lantz's solution was to propose a government program that would buy back from homesteaders land the government gave them only a couple of decades before, aggregate it and let it go back to grass, a grazing commons, federally owned. Lantz had the ear of his former boss, M. L. Wilson, who had quit Montana to join the administration of Franklin Roosevelt. Wilson, in turn, had the ears of such as Henry Wallace, FDR's famously progressive agriculture secretary and Rexford Tugwell, known widely by the press as "Rex the Red." All of these men were in precisely the same position as are Barack Obama's people as I write this. Newly in office they were desperate for ideas that were instantly doable, a need captured by the term "shovel ready" in the stimulus plans of today's parallel political universe. Lantz's idea became the Resettlement Administration and the Bankhead-Jones Farm Tenant Act. The former bought back from busted homesteaders 11.3 million acres during the Great Depression, most of which aggregated under what was to become the Bureau of Land Management, which today superintends an area of public lands in the West the size of France. A subset of acquisitions under the latter bought up 3.75 million acres managed now by the U.S. Forest Service as our nation's national grasslands, the locations of many of the scenes on the following pages. Coupled with the Taylor Grazing Act, this legislation formed the basis of conservation in the West, a statement that no doubt goes down hard with those of us who fight for the integrity of arid places. The Grazing Act, the BLM and Forest Service all are regarded widely as the enemies of conservation. How can this be?

There can be no doubt about the sincerity of Franklin Roosevelt's commitment to conservation; he was among our best, particularly as conservation was fomented by Harold Ickes, his legendary secretary of interior. They meant well, as did Lantz. The devil was in the details, sweeping programs of reform hatched hastily in a time of great need. Yet not all was in the details. As Roosevelt became distracted by the towering challenges of the war and even more so after he died, ranchers and other vested interests were able to undermine a couple of key provisions of the Taylor Grazing Act.

First, both Lantz and the law that set up these grasslands specified they were to be an unfenced commons, in which a community of ranchers had a right to graze a certain number of animals. The BLM and Forest Service later acquiesced to rancher pressure to smash the commons into allotments, a certain acreage granted to each rancher. This was a de-facto privatization of these hard-won public lands.

The result has been overgrazing and abuse on half a continent. So where is the progress? You can see it for yourself on the following pages. These lands are covered in grass. They are not pristine, but by no means are they as savaged as during the Dust Bowl. Further they are still, nominally at least, public lands, and this is a time of great optimism in our nation. If we read this legacy correctly, it can help tell us where we may go. And if we can read it correctly it can caution against haste, can make us as patient and as unswerving as the place itself, so we might insert no devil in the details.

Besides, this is not a story plucked from the Great Depression and dusted off. It has not been dormant. I have been following it for twenty years and can now report some key developments. Two decades ago, you could kick up a handful of people throughout the plains thinking about and acting on the coming resurrection of the plains. These are the people educated by grass, and their numbers have risen exponentially since. Their methods have matured, and some of their thinking is now embedded in both the BLM and Forest Service. We as a people have come to respect the grasslands as something more than wasteland, "flyover country."

Further, we have come to once again envision the grasslands as not a grazing commons but as wild.

The second thing I have to report is that this group of people have conceived of the creation of a 3 million acre bison preserve, part of it in Phillips County, Montana, and have begun buying up ranches to give them control of the very acres bought with government checks issued by Henry Lantz. The preserve is already a third of the way home, in that at its center lies the Charles M. Russell National Wildlife Refuge, 1.1 million acres of public lands along the Missouri River set aside by Franklin Roosevelt. The fact that a significant herd of native, free-roaming bison began inhabiting former ranches bordering the refuge a couple of years ago is at once testimony to his legacy and a charge to us on how we might carry it forward from these optimistic times.

But all of this requires vision, both figurative and literal. Catlin began imagining the place in vast terms after he had painted it. To see is to believe, so it is fortunate now we can include Don Kirby in Catlin's lineage. His vision, too, is timely.

Richard Manning

FOR MY WIFE, JOAN GENTRY, WITH LOVE

Twotrack, Sandhills, Halsey, Nebraska, 2006

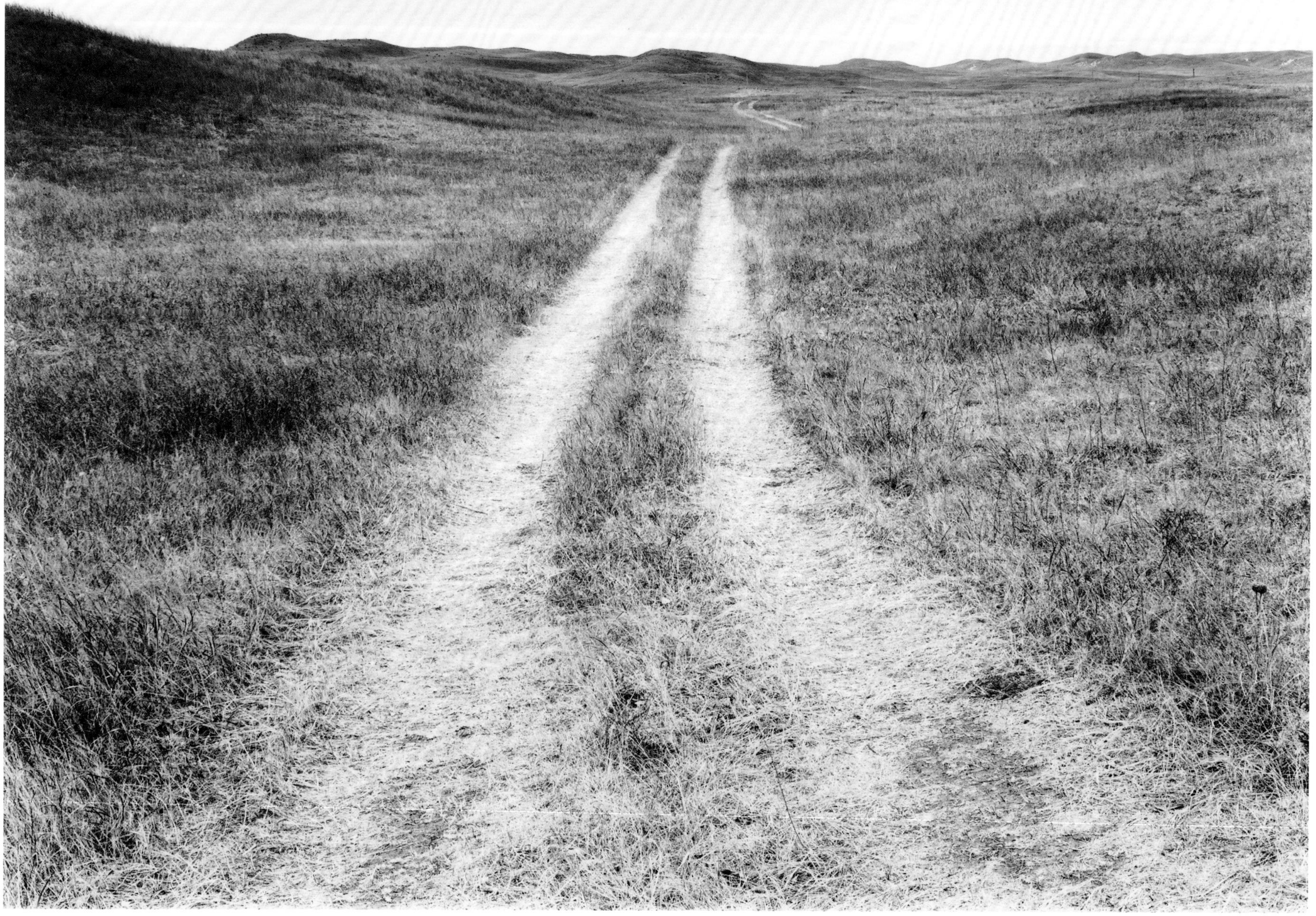

Hayranch, Route 97, Sandhills, Nebraska, 2006

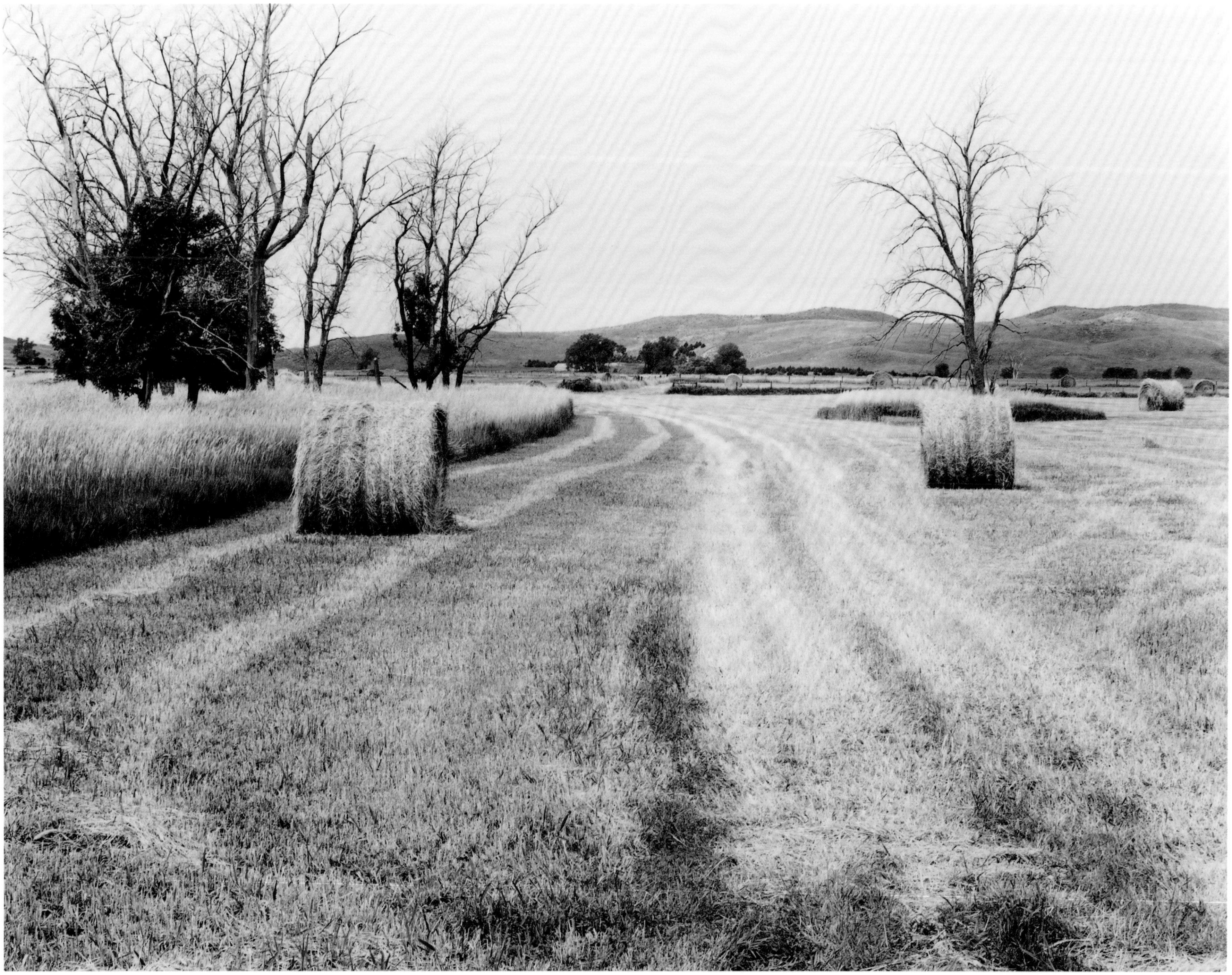

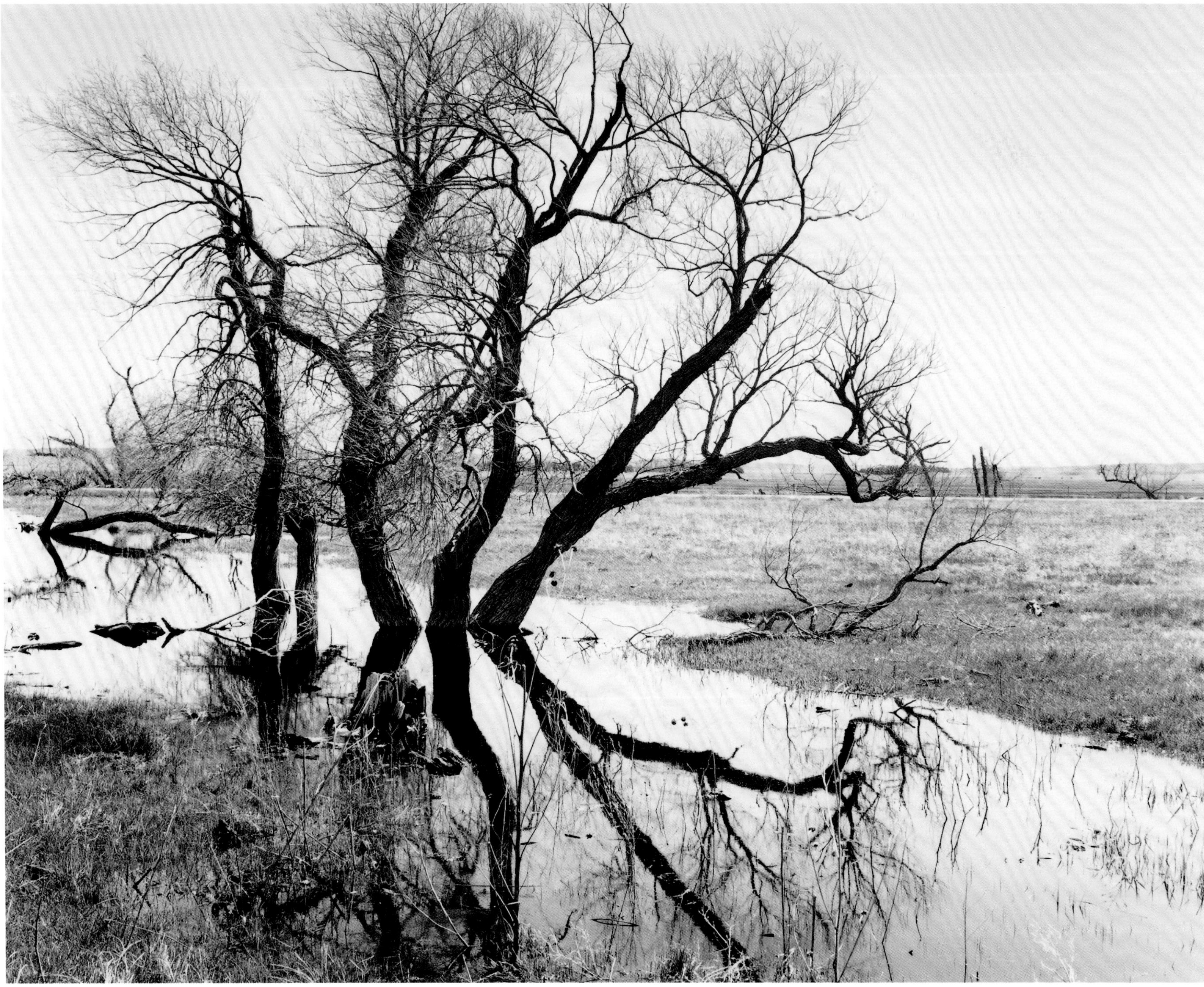

Reflections, Eatinger Ranch, Sandhills, Nebraska, 2007

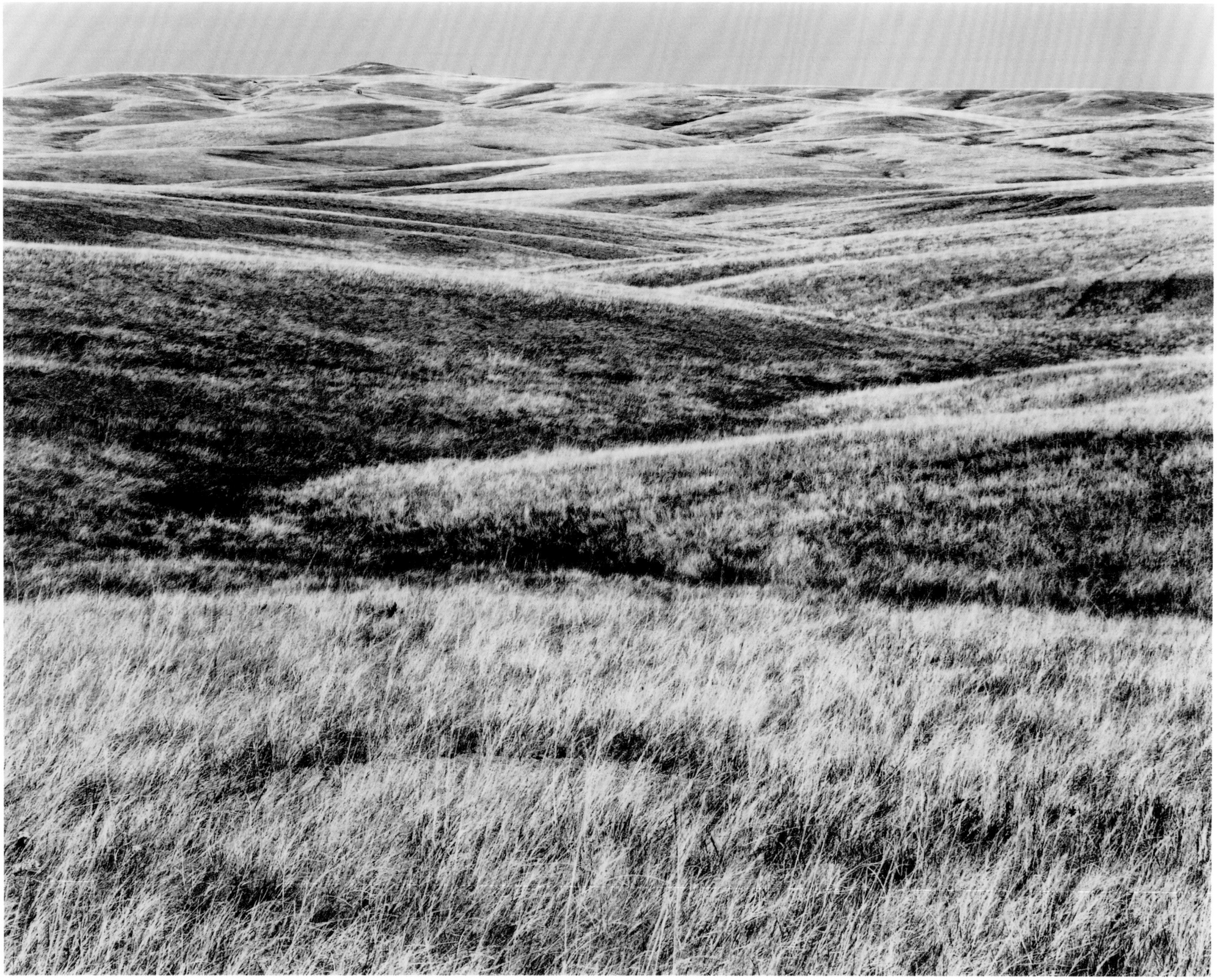

Horsehead Creek Drainage, Oglala National Grasslands, Nebraska, 2006

North Loup River, Miles Ranch, Sandhills, Nebraska, 2007

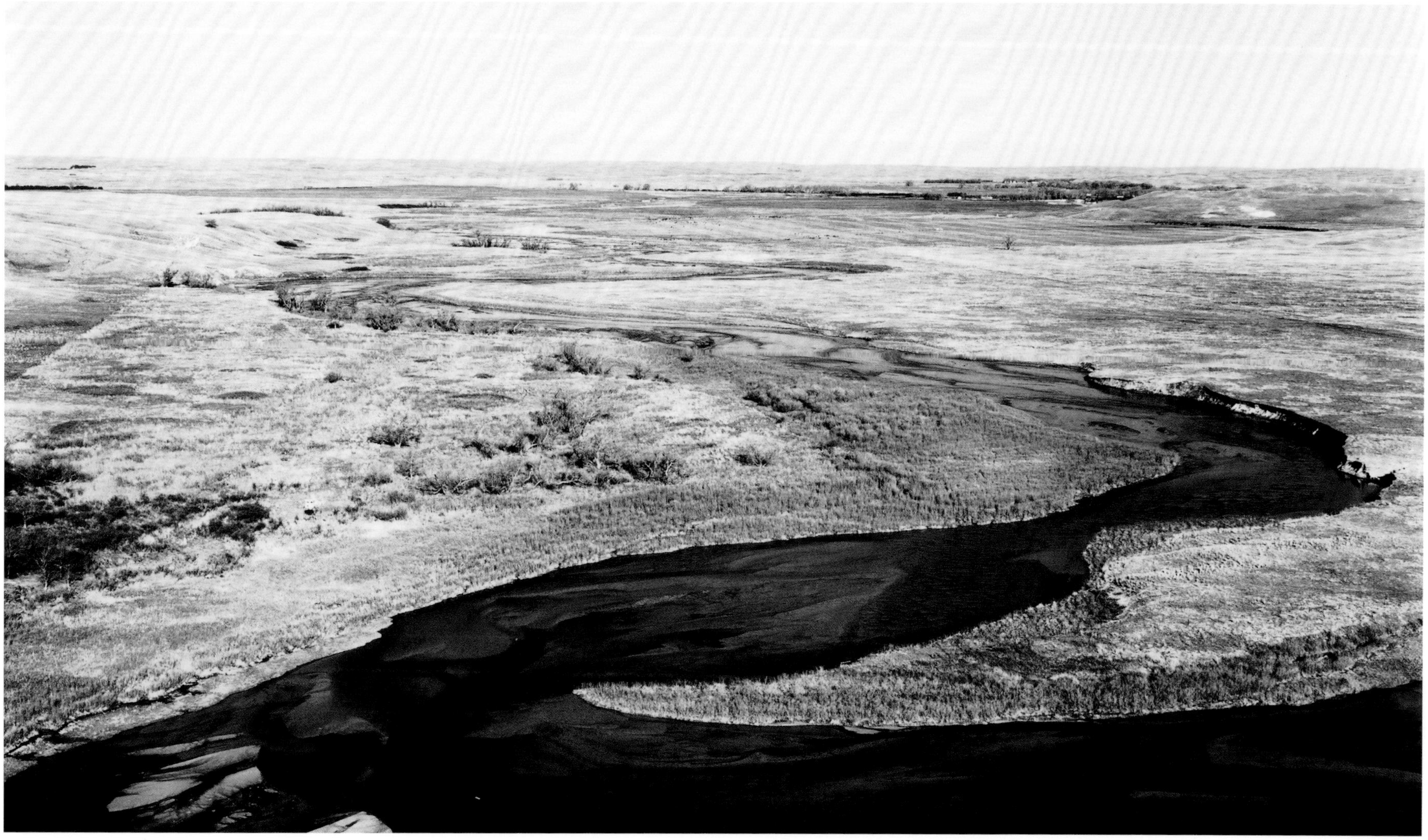

Line Fence, Buffalo Gap National Grasslands, South Dakota, 2006

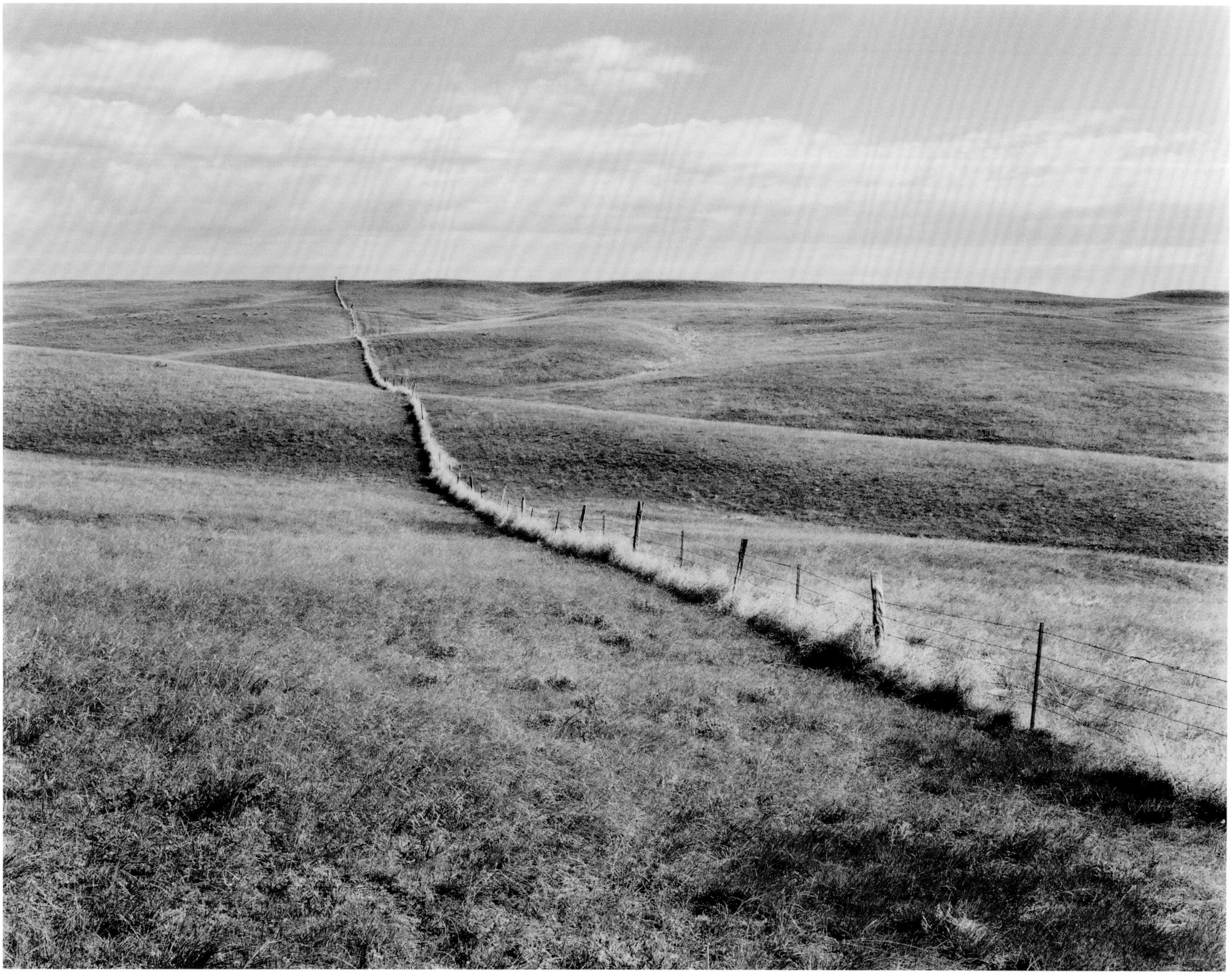

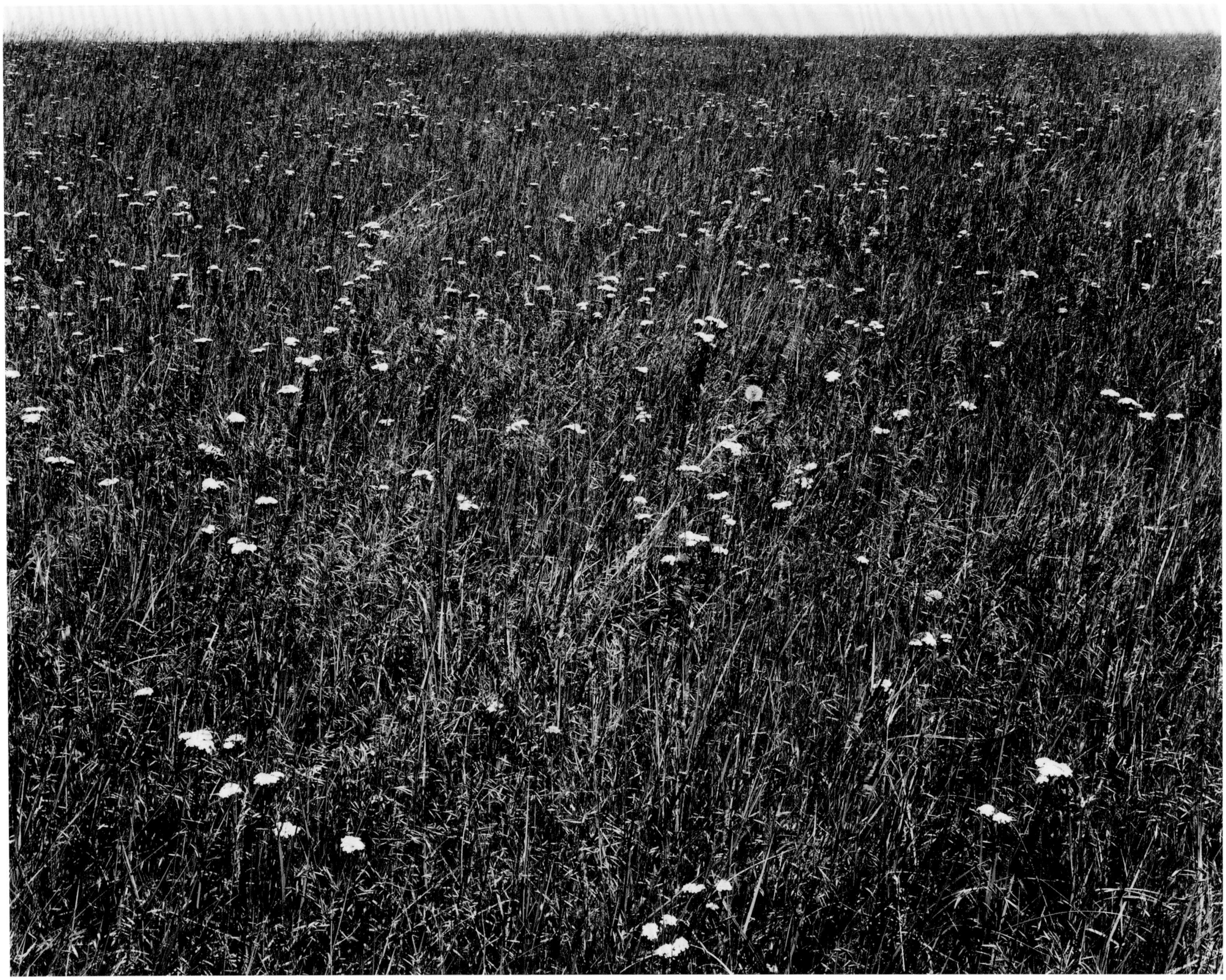

Fort Pierre National Grasslands, South Dakota, 2008

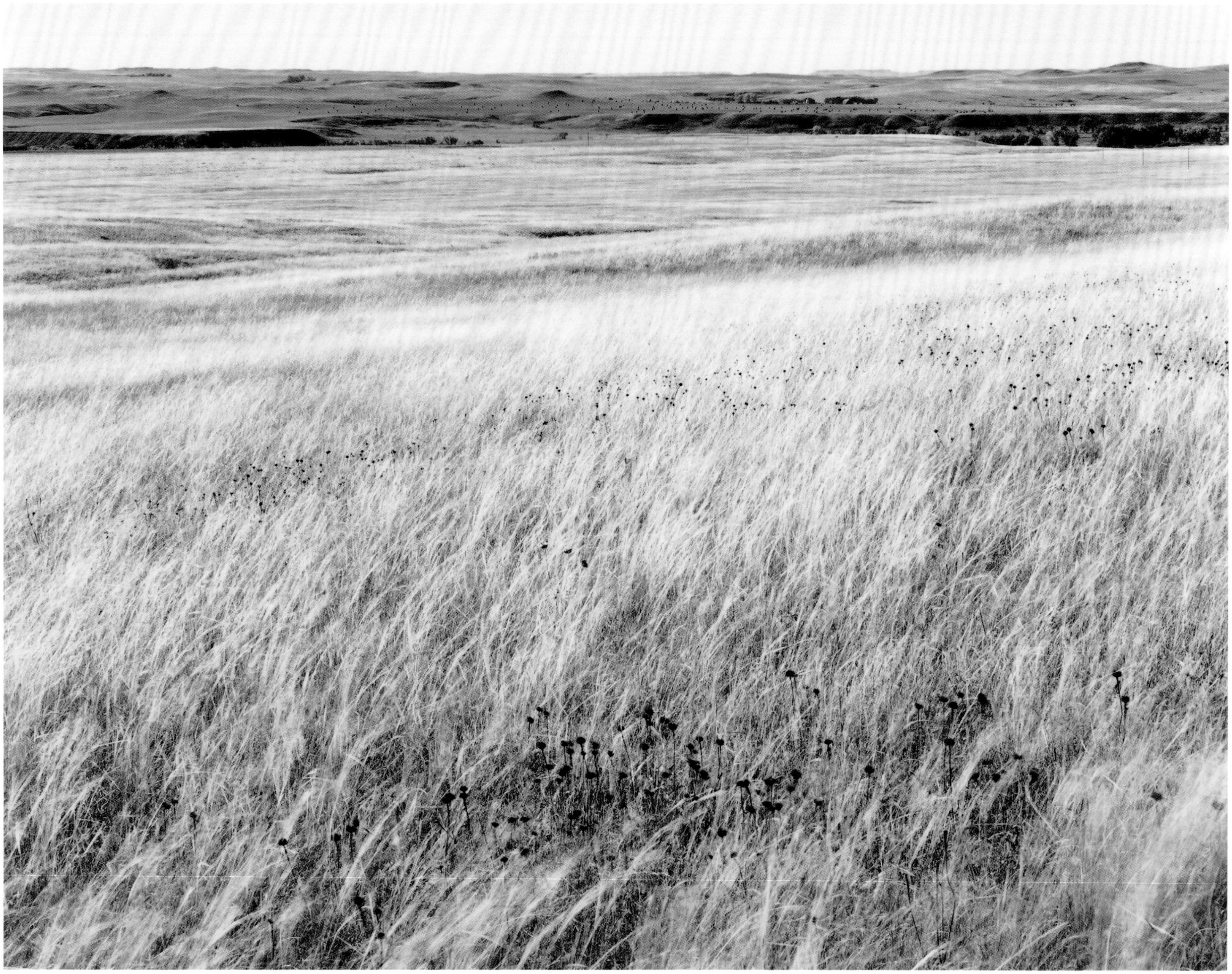

Grand River National Grasslands, South Dakota, 2007

Lone Butte, Buffalo Gap National Grasslands, South Dakota, 2006

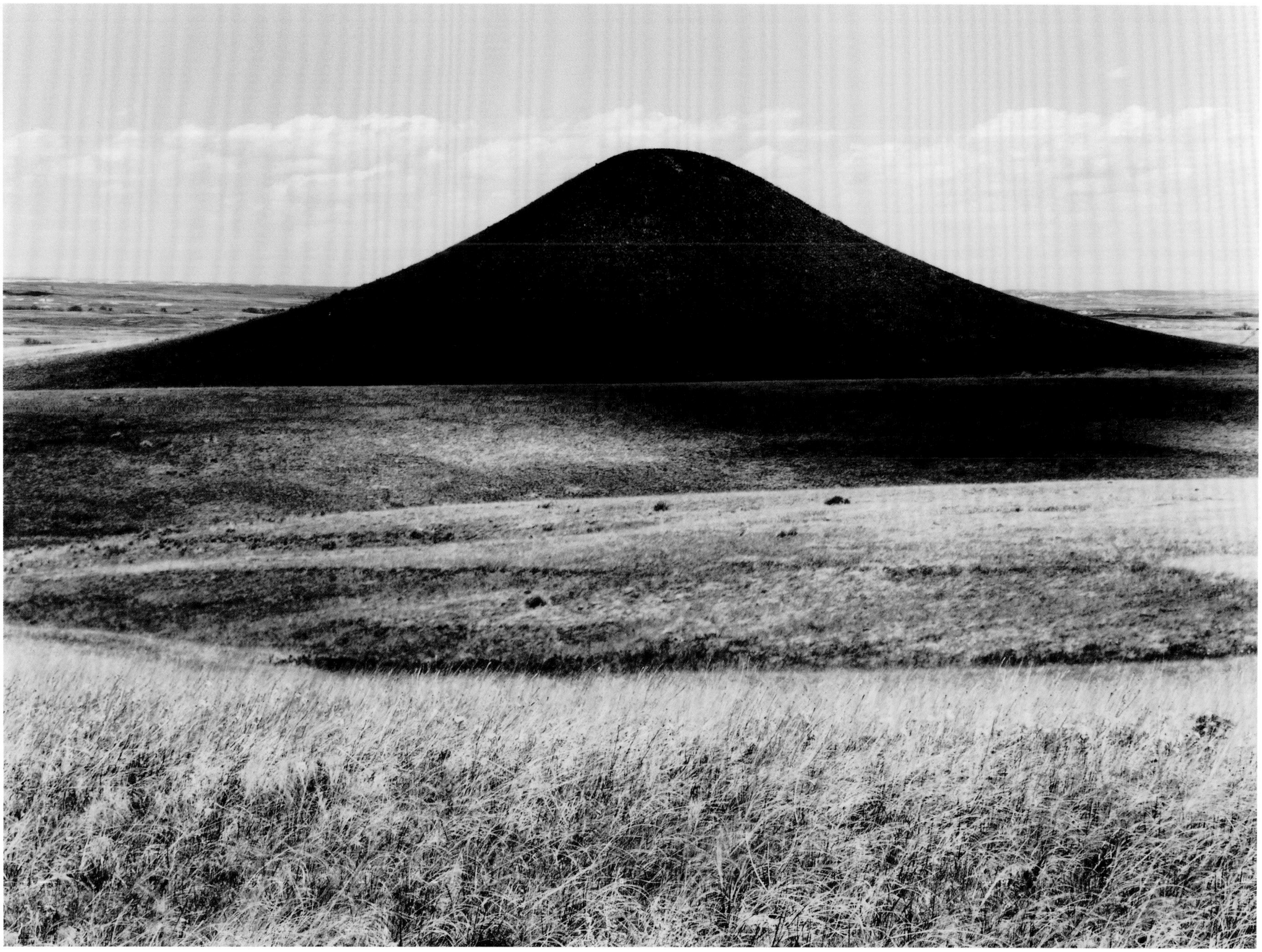

Pond, Vicinity Hay Canyon Butte, Buffalo Gap National Grasslands, South Dakota, 2006

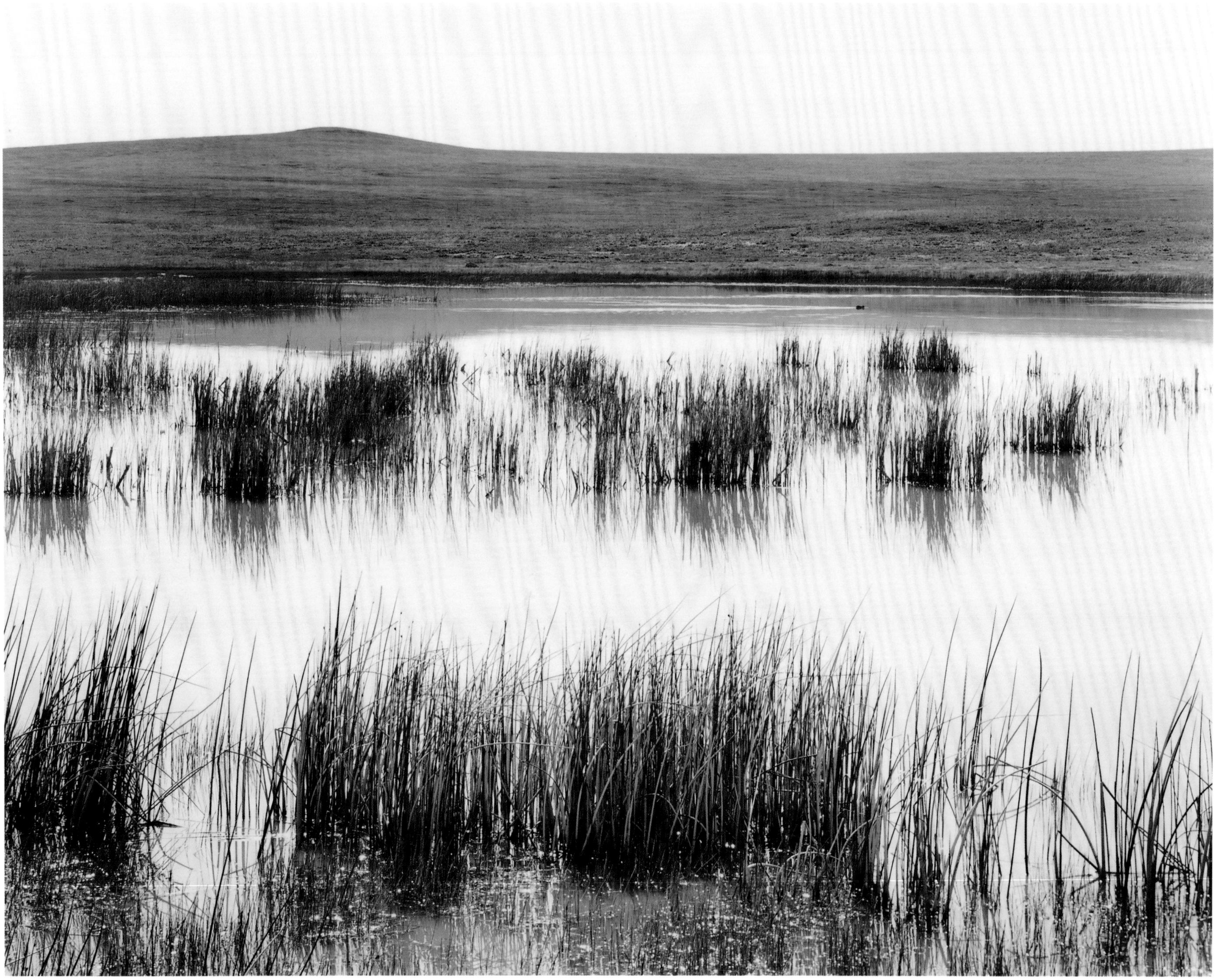

Grand River National Grasslands, South Dakota, 2007

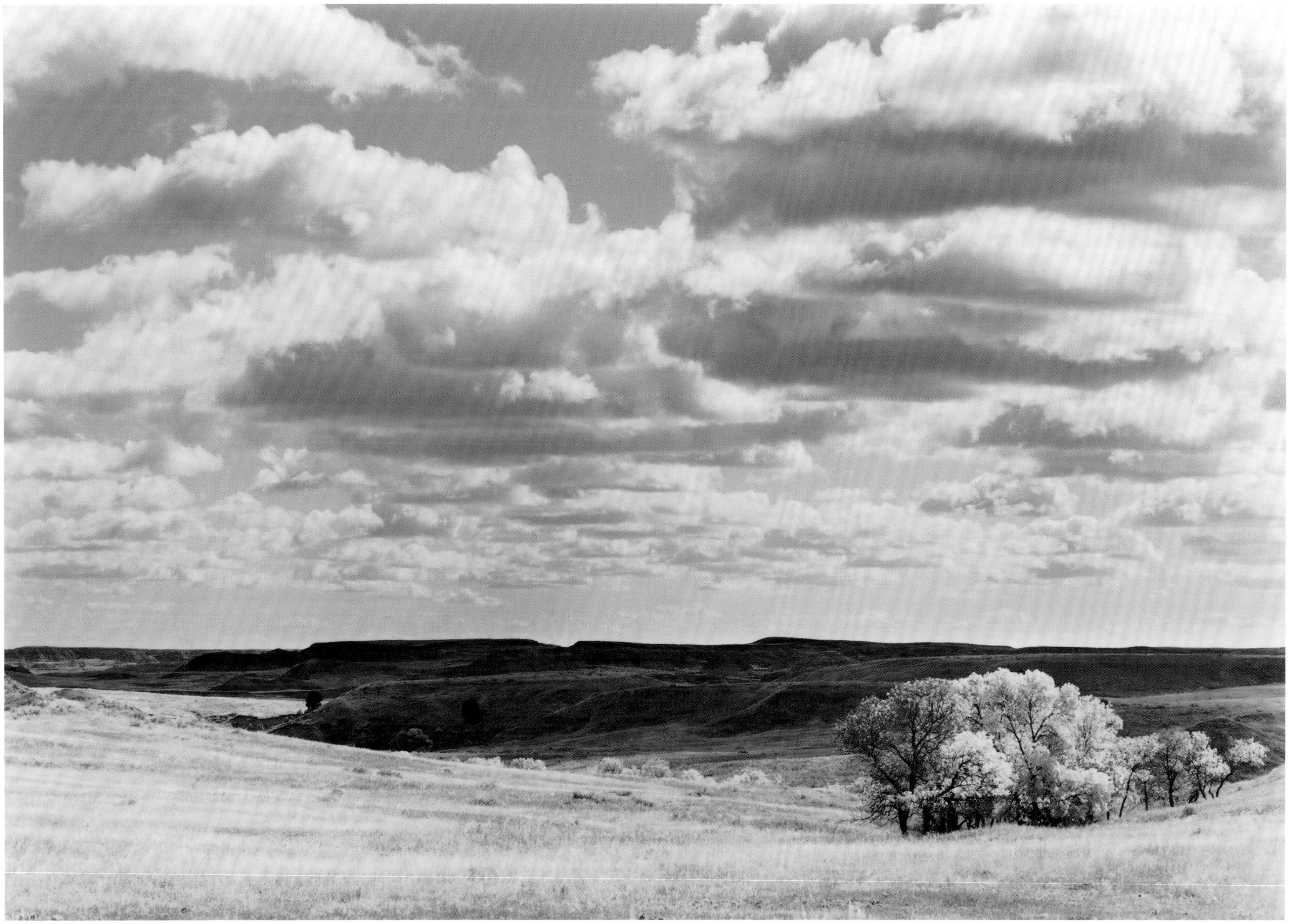

Approaching Storm, Fort Pierre National Grasslands, South Dakota, 2008

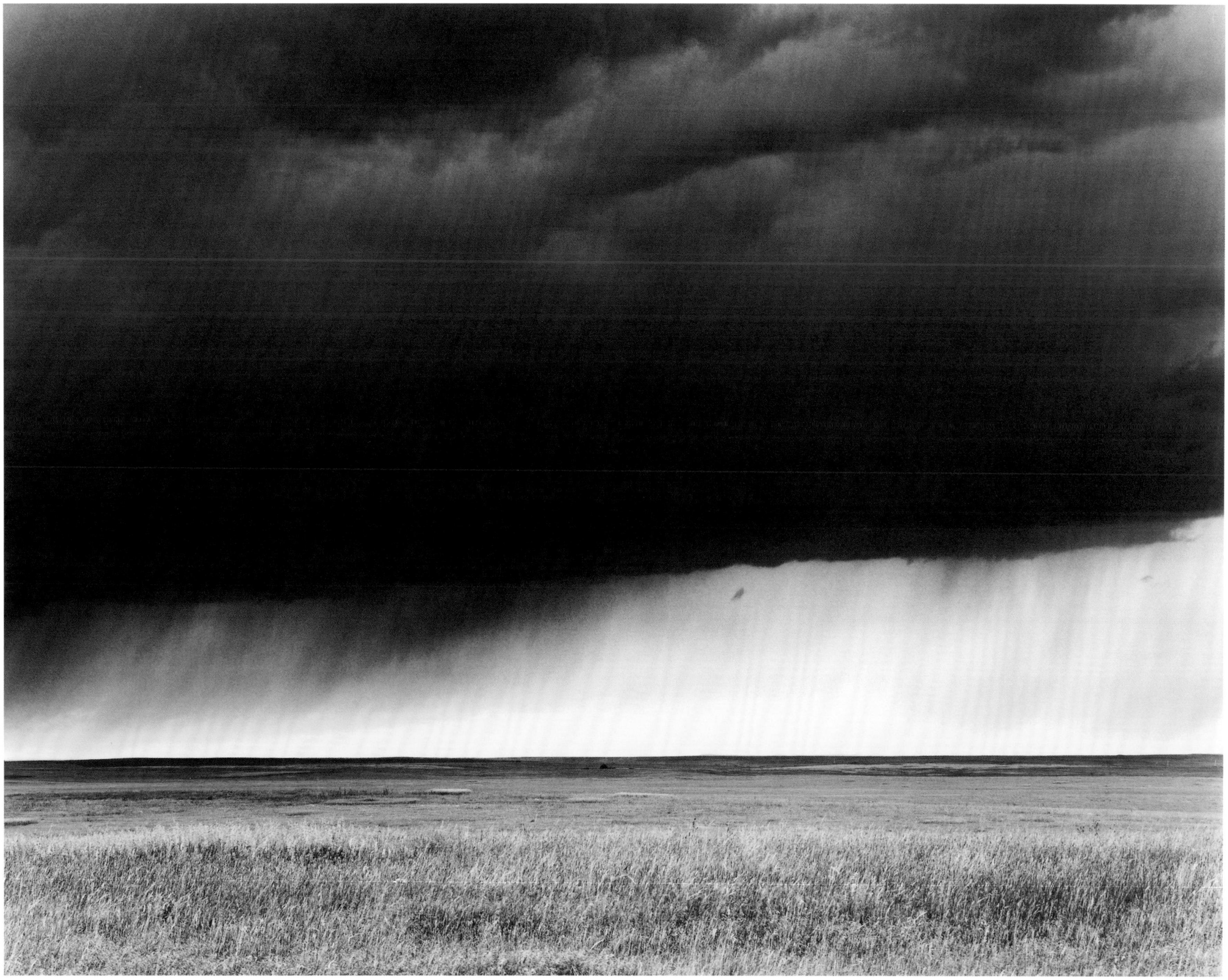

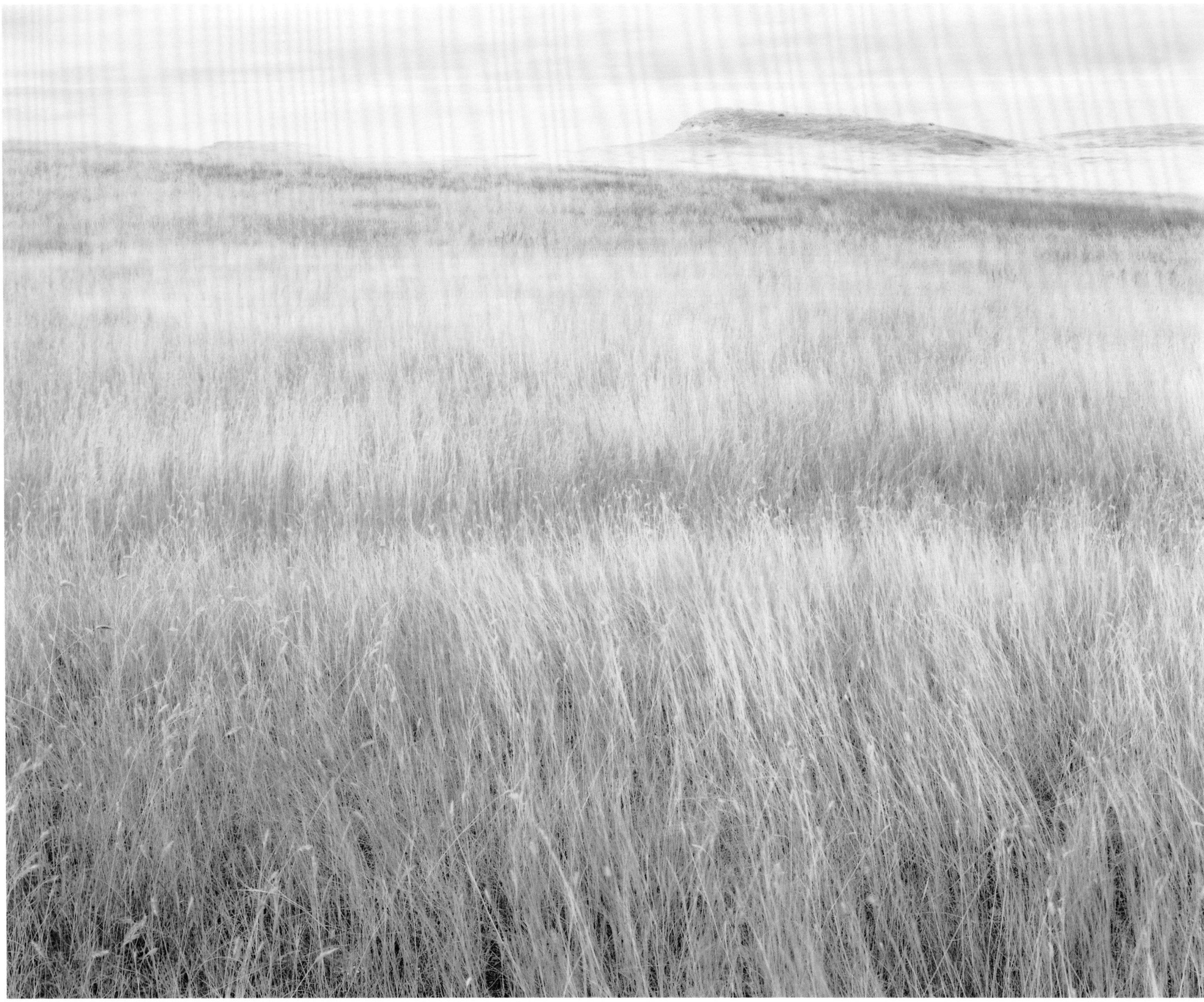

Little Missouri National Grasslands, North Dakota, 2007

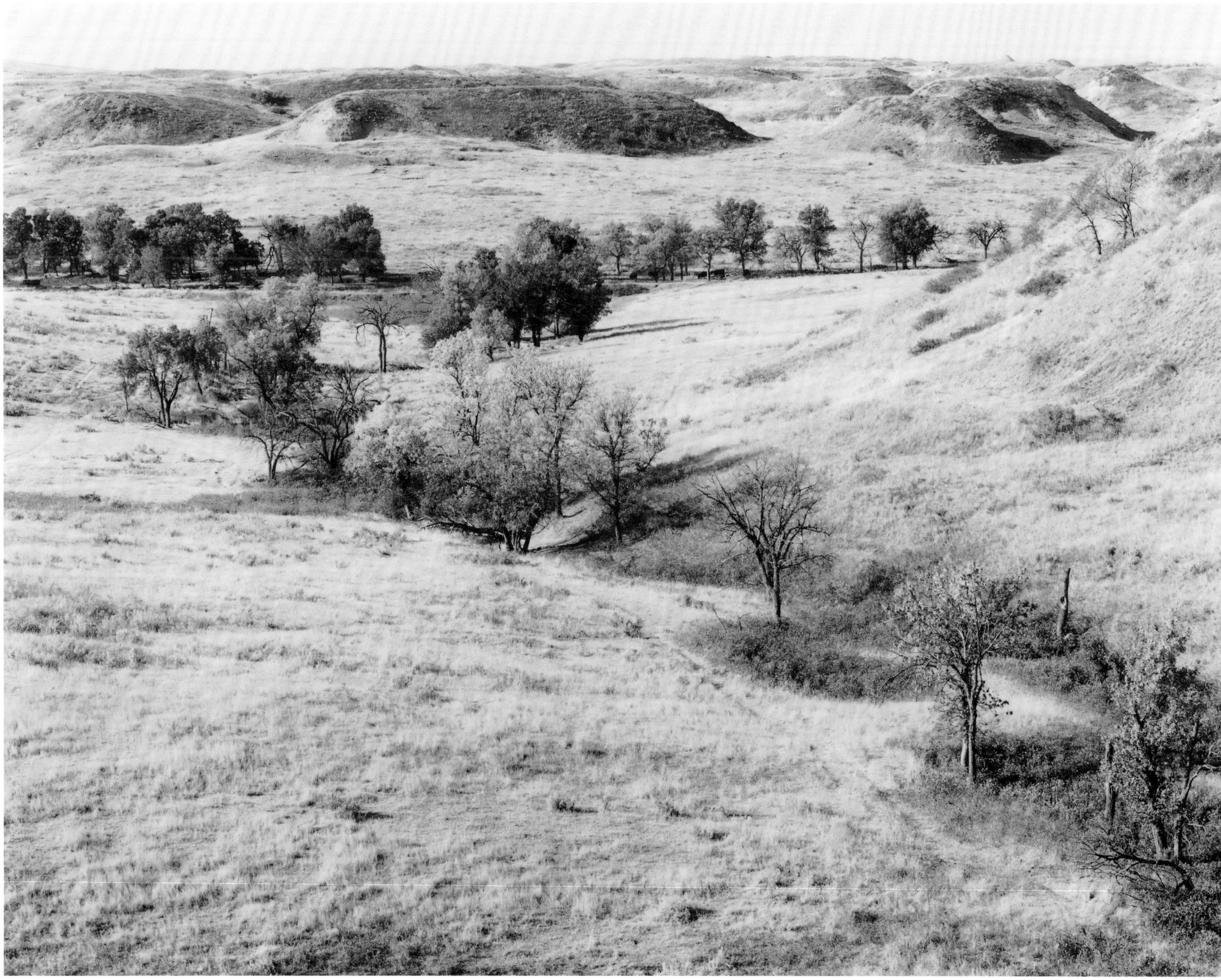

Little Missouri National Grasslands, North Dakota, 2007

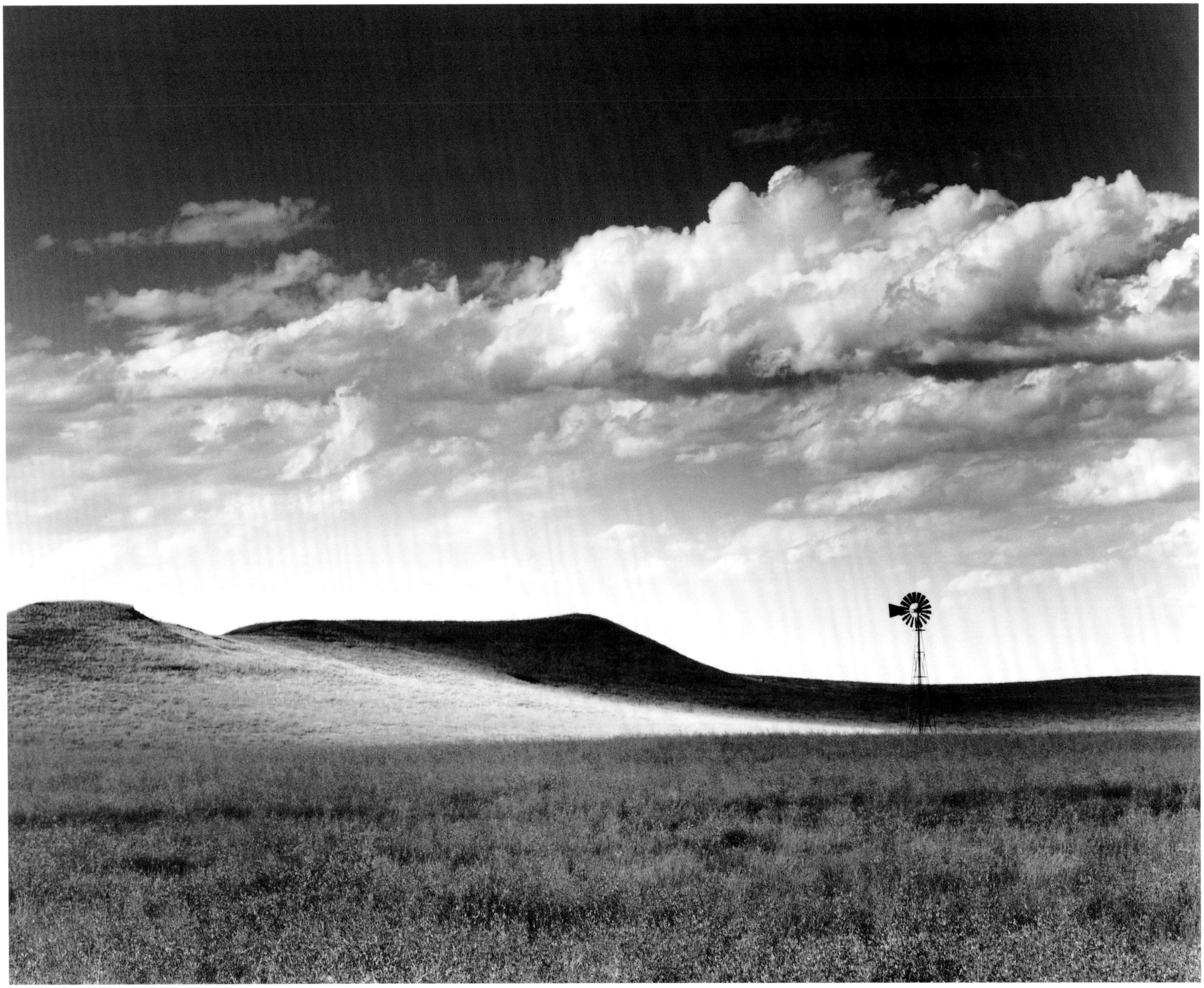

Cedar River National Grasslands, North Dakota, 2008

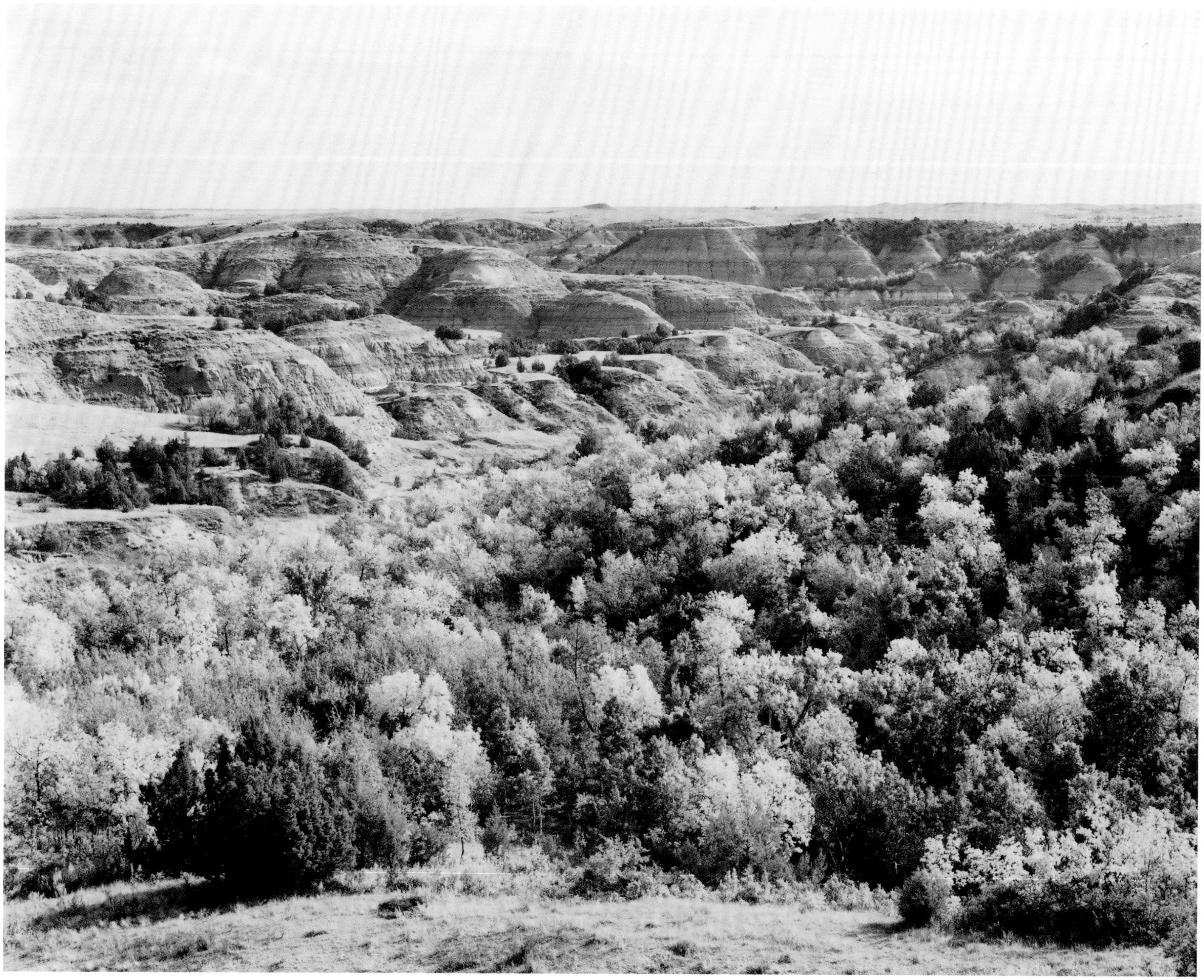

Little Missouri National Grasslands, North Dakota, 2007

Little Missouri National Grasslands, North Dakota, 2007

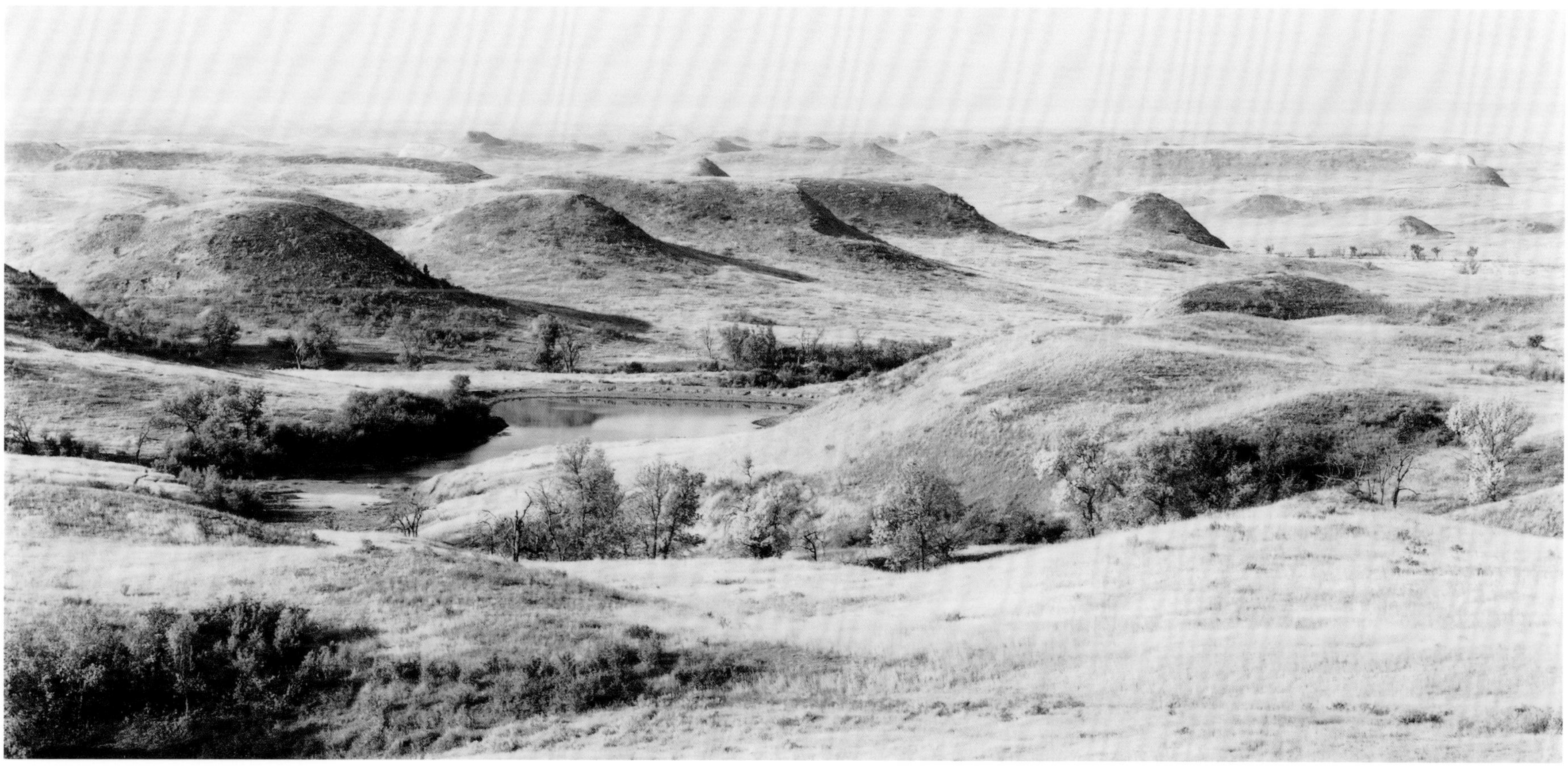

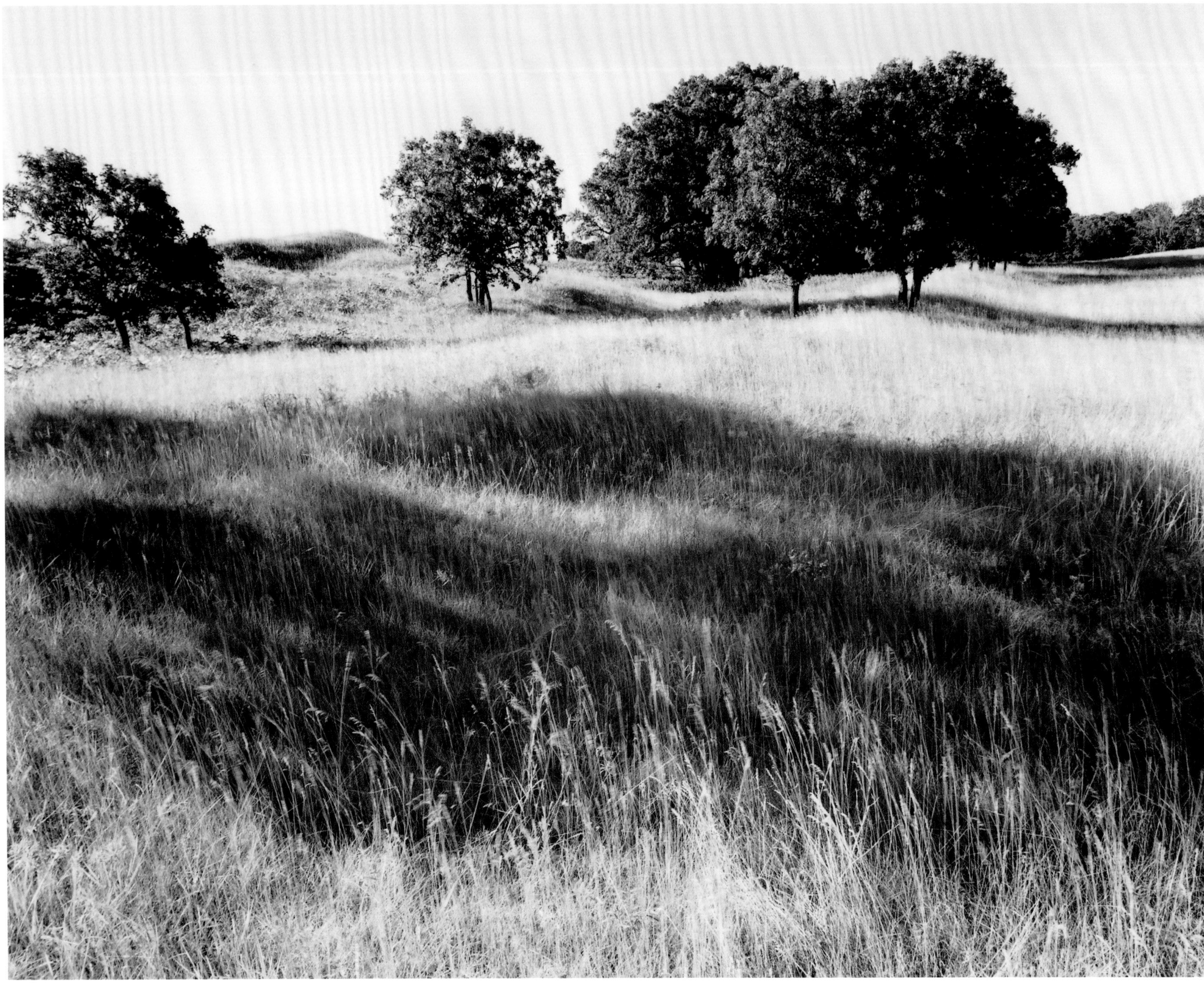

Sheyenne National Grasslands, North Dakota, 2008

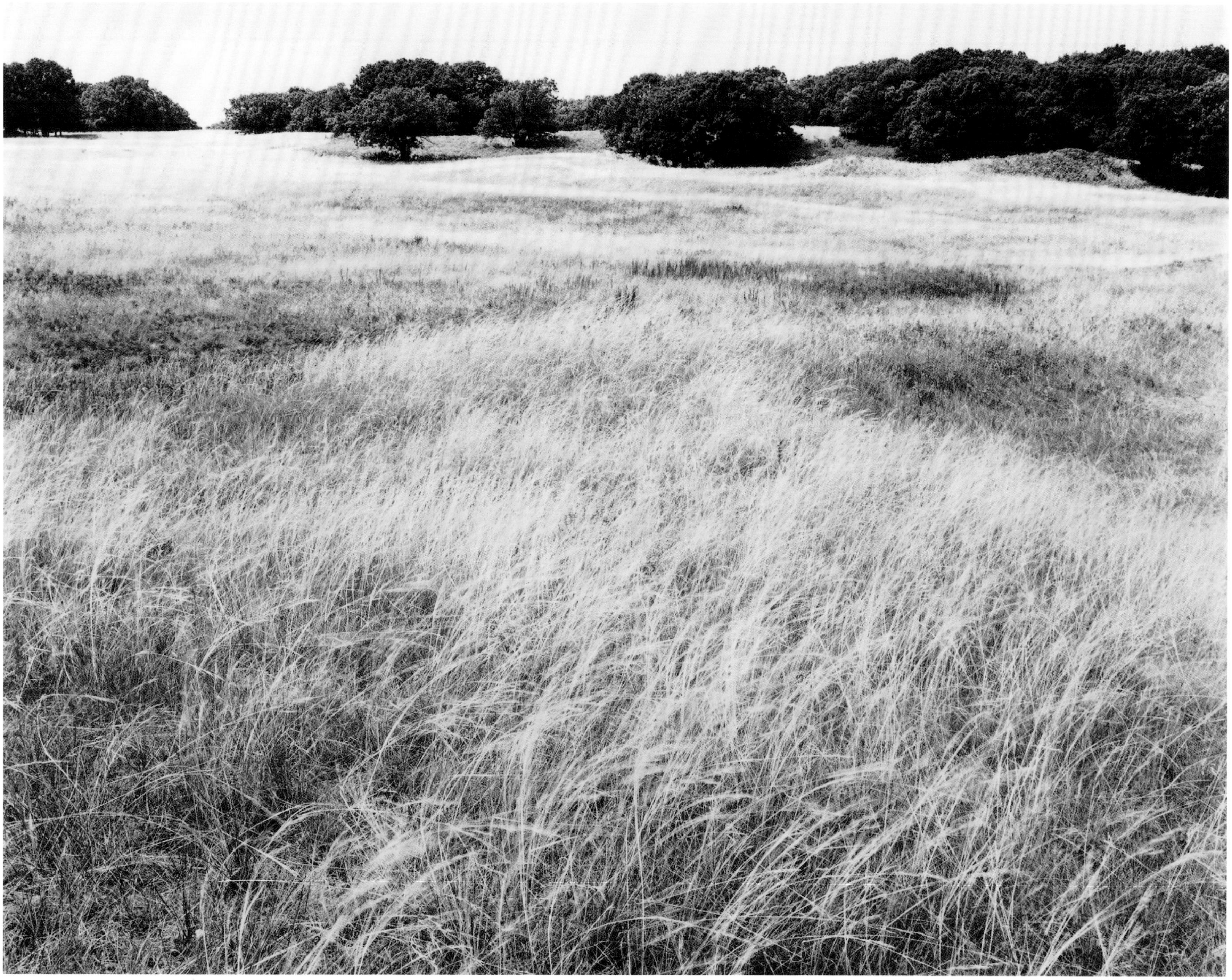

Sheyenne National Grasslands, North Dakota, 2008

Lone Tree, Cimarron National Grasslands, Kansas, 2007

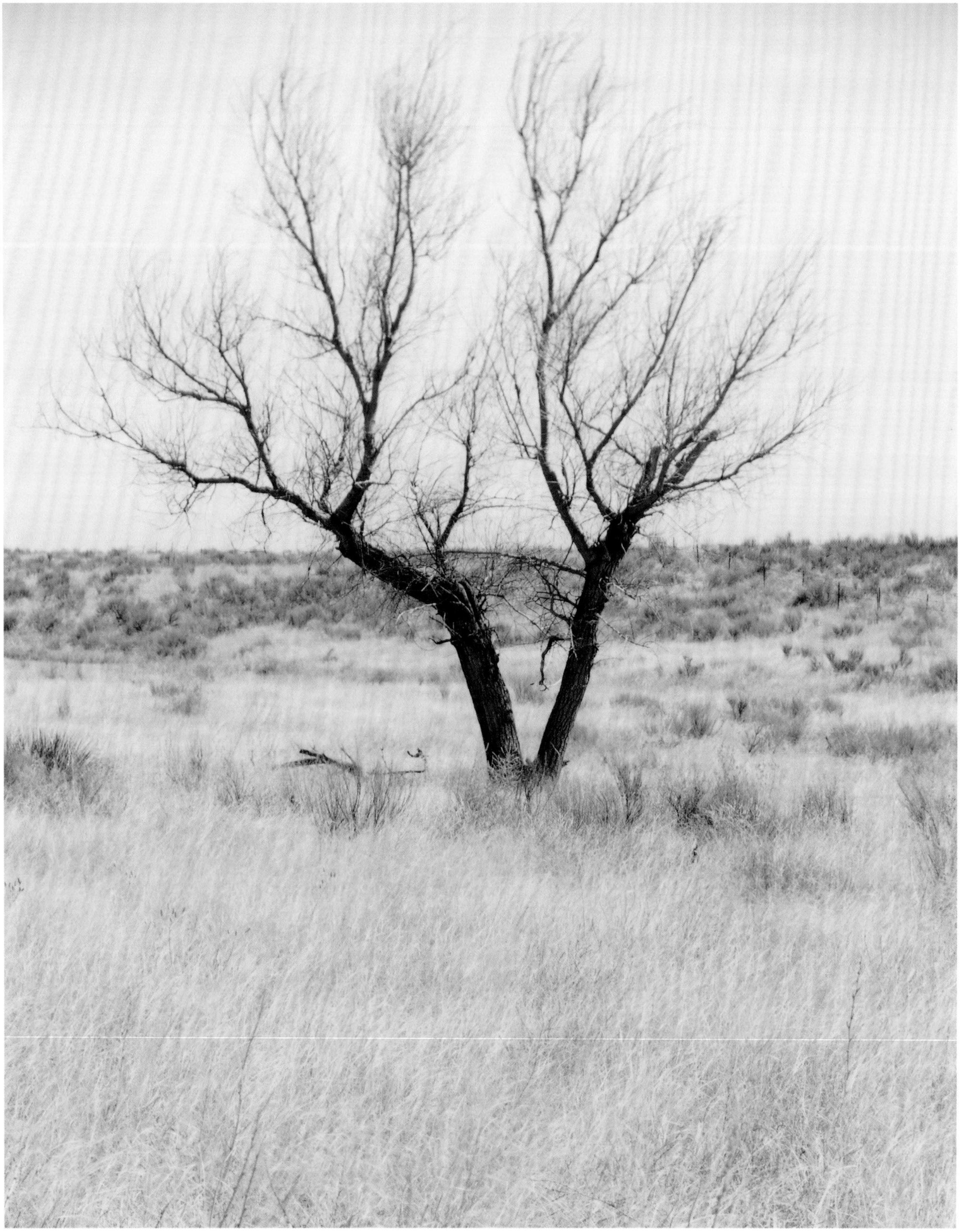

Controlled Burn, Cimarron National Grasslands, Kansas, 2007

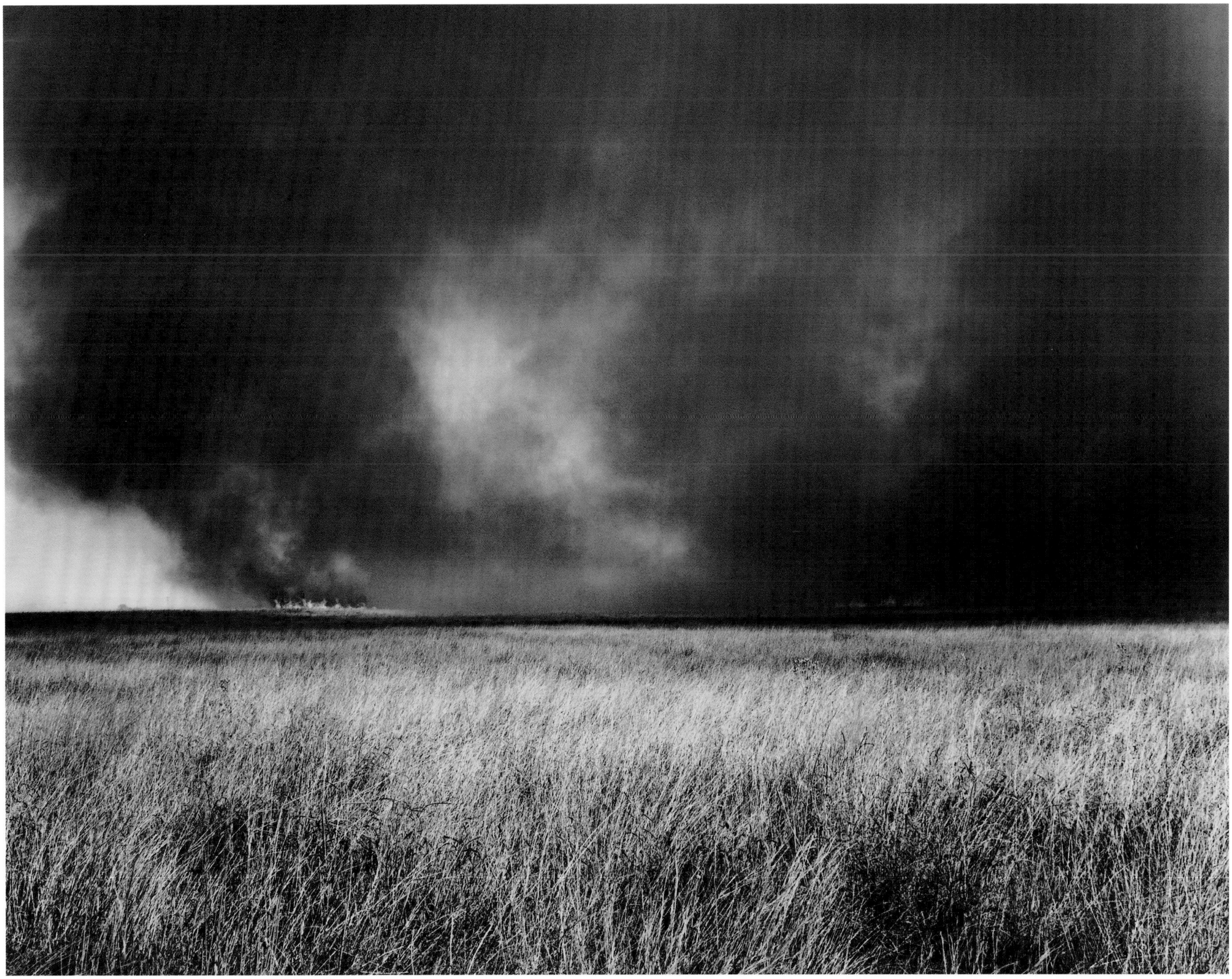

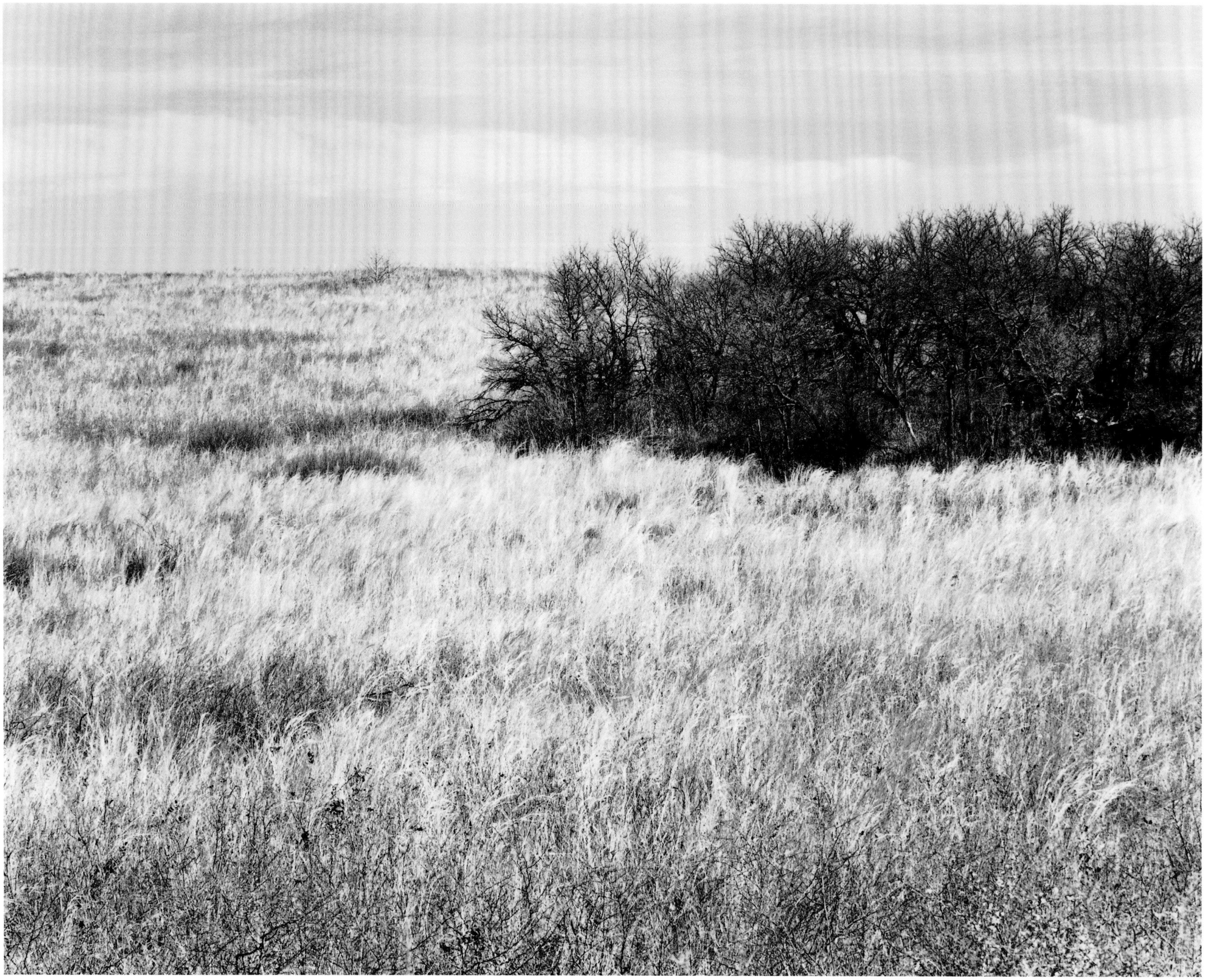

Black Kettle National Grasslands, Oklahoma, 2008

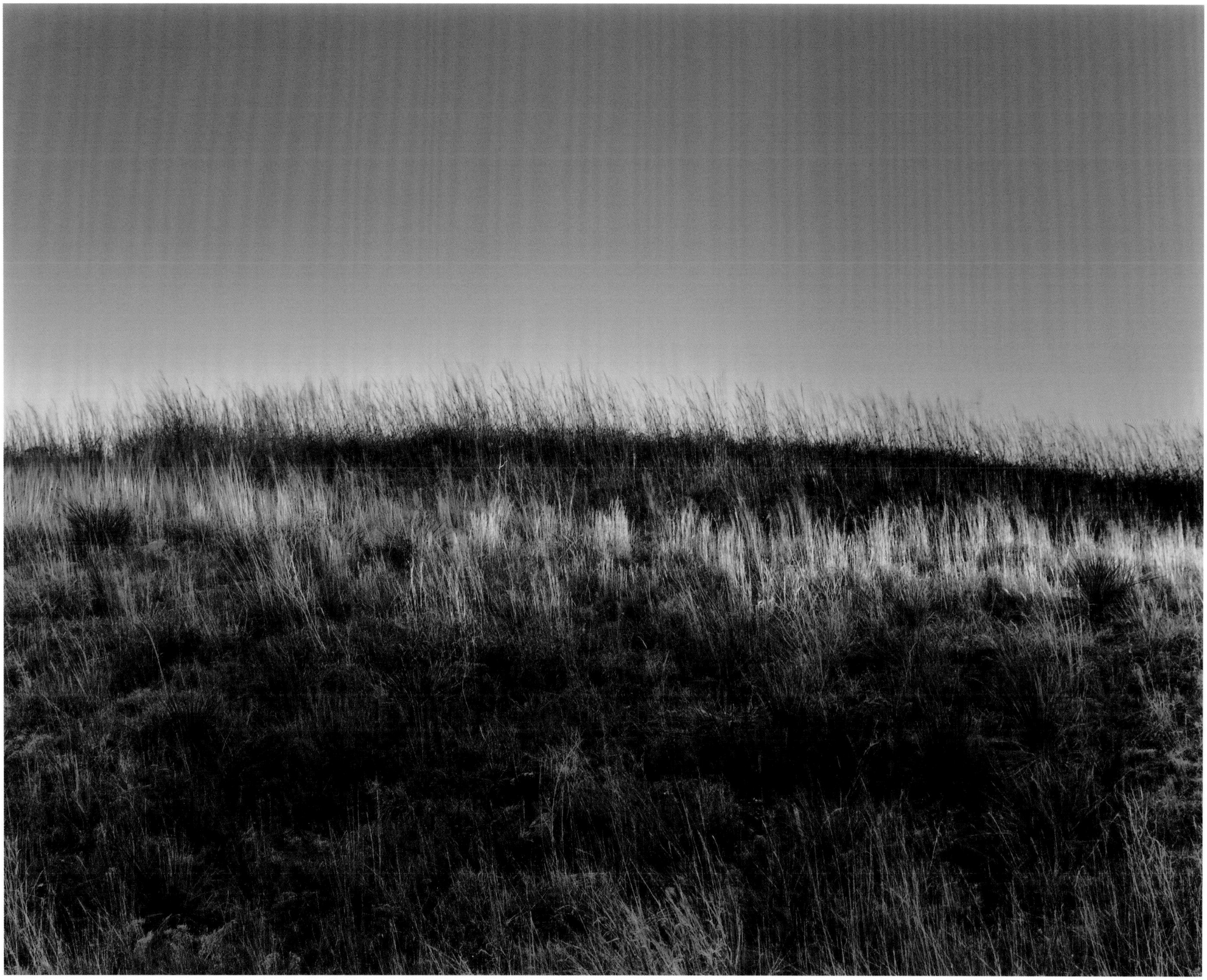

Last Ray, McClellan Creek National Grasslands, Oklahoma, 2008

Black Kettle National Grasslands, Oklahoma, 2008

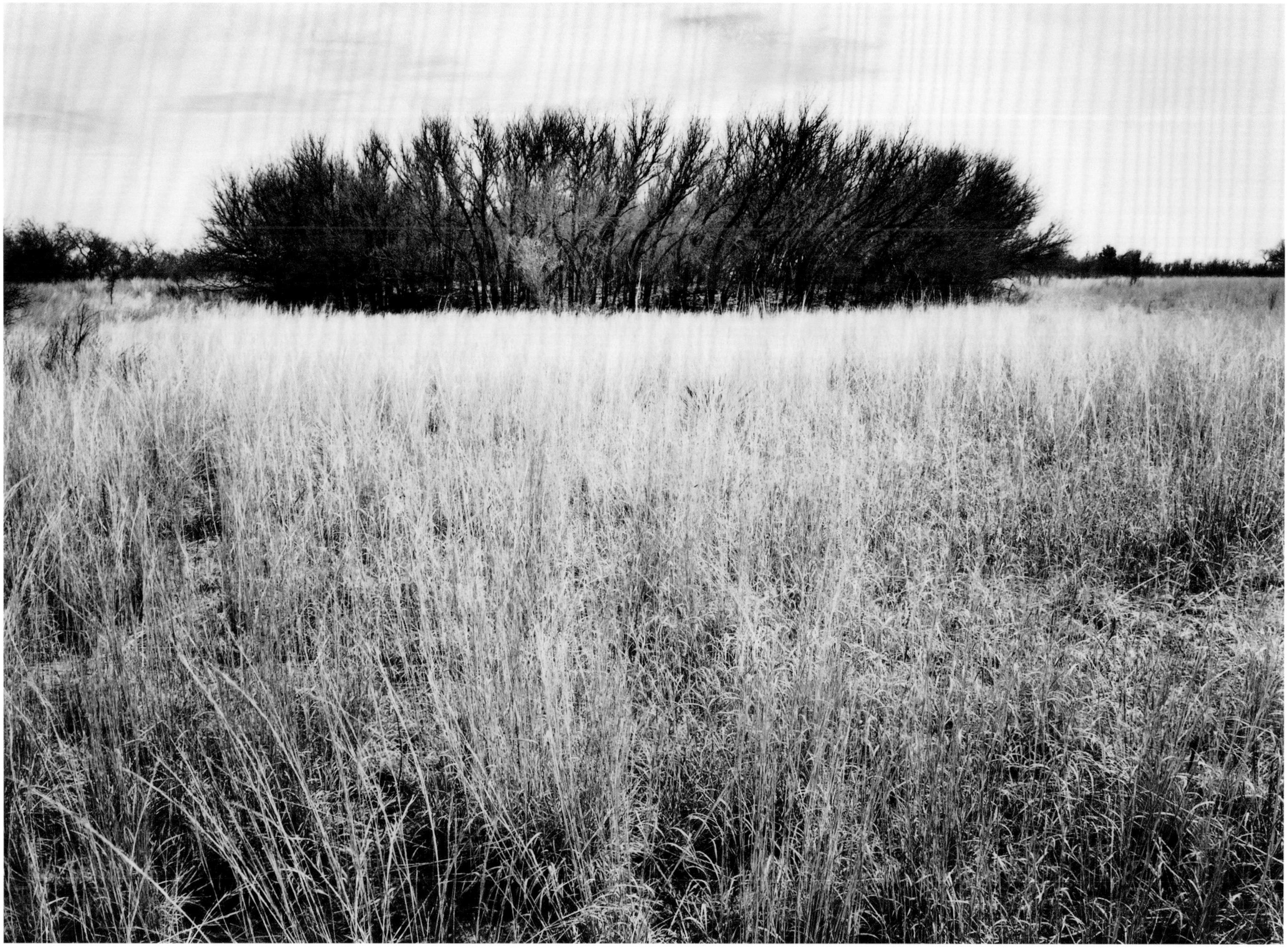

Approaching Storm, Rita Blanca National Grasslands, Texas, 2006

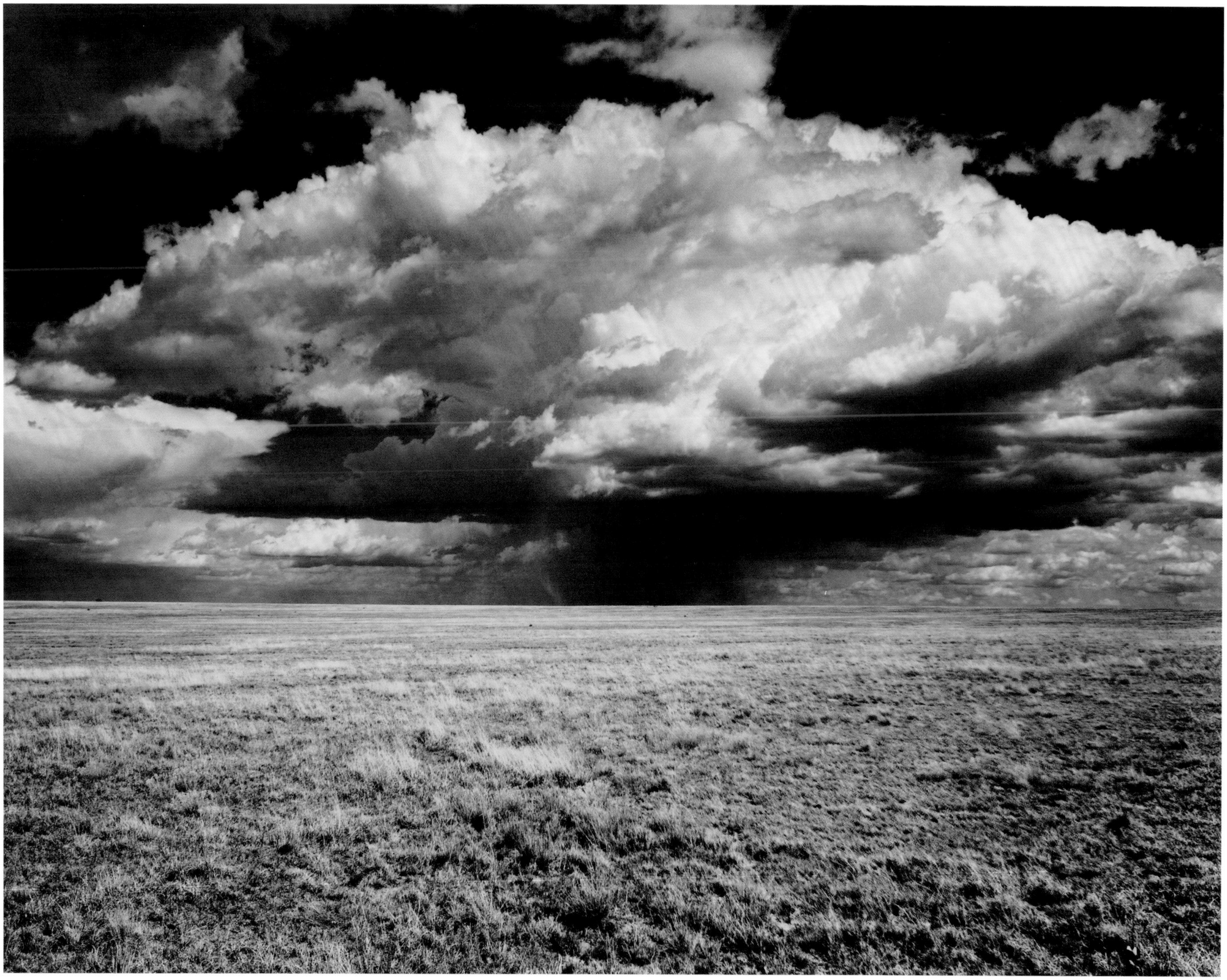

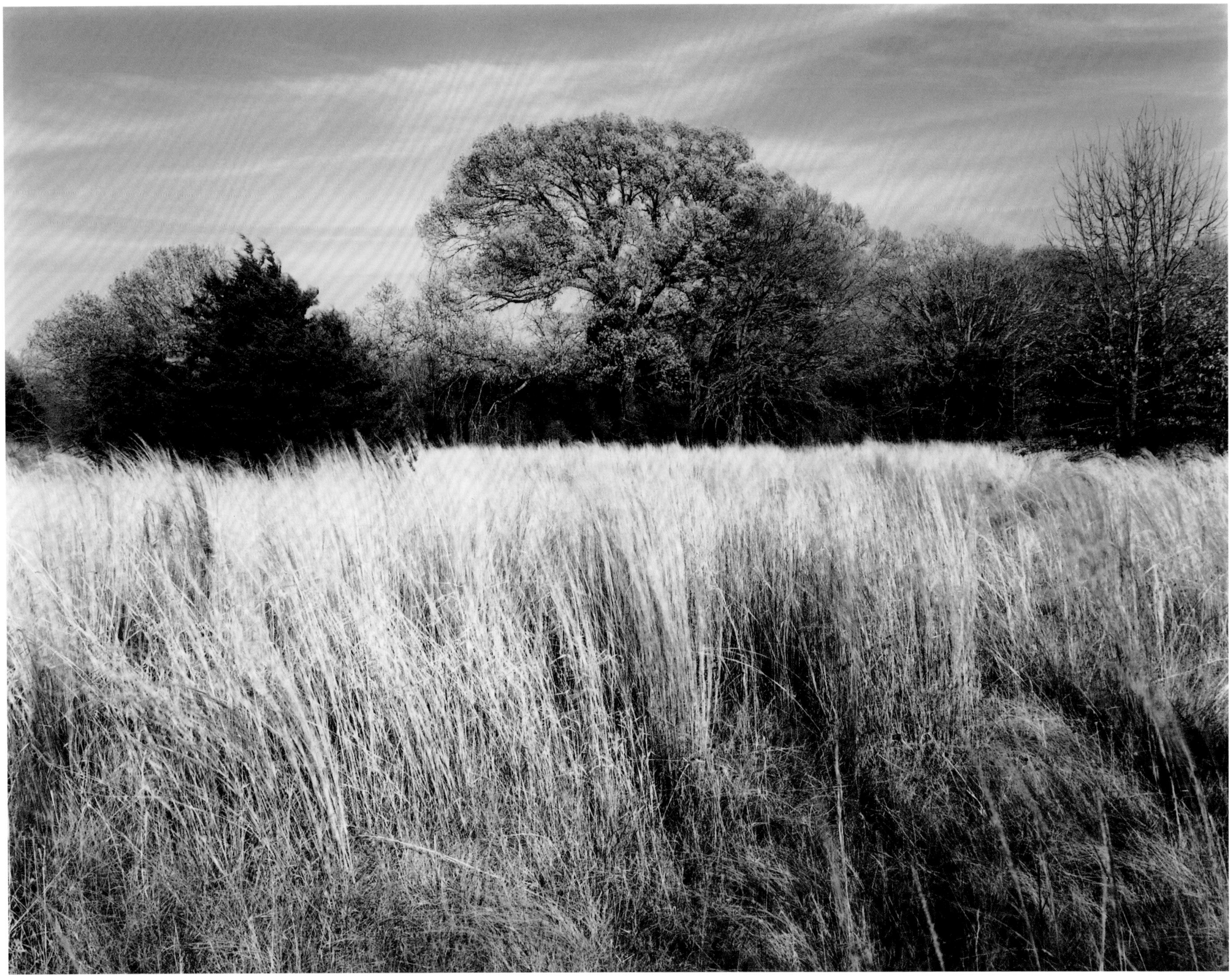

Caddo National Grasslands, Texas, 2008

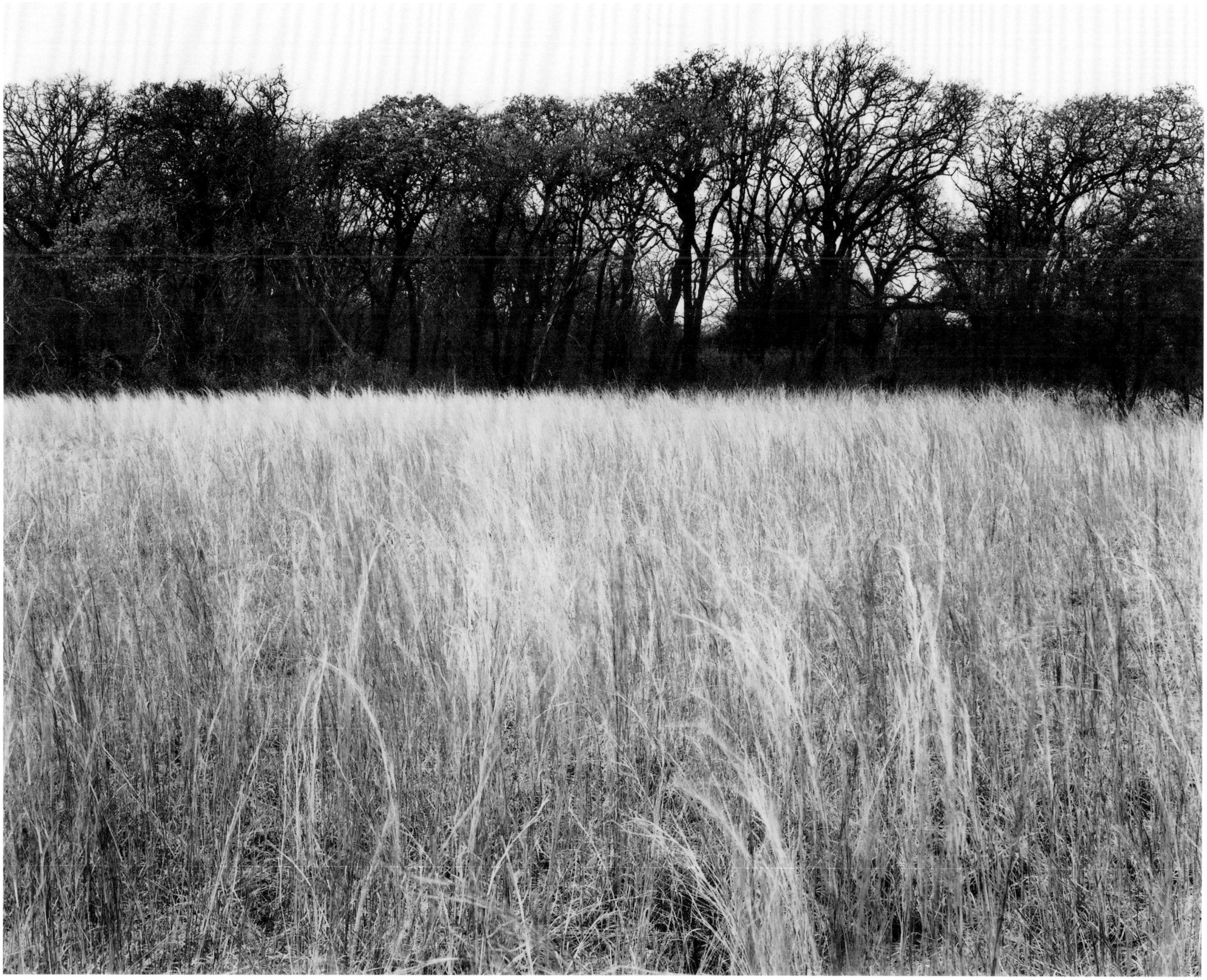

LBJ National Grasslands, Texas, 2008

Soap Weed, Kiowa National Grasslands, New Mexico, 2006

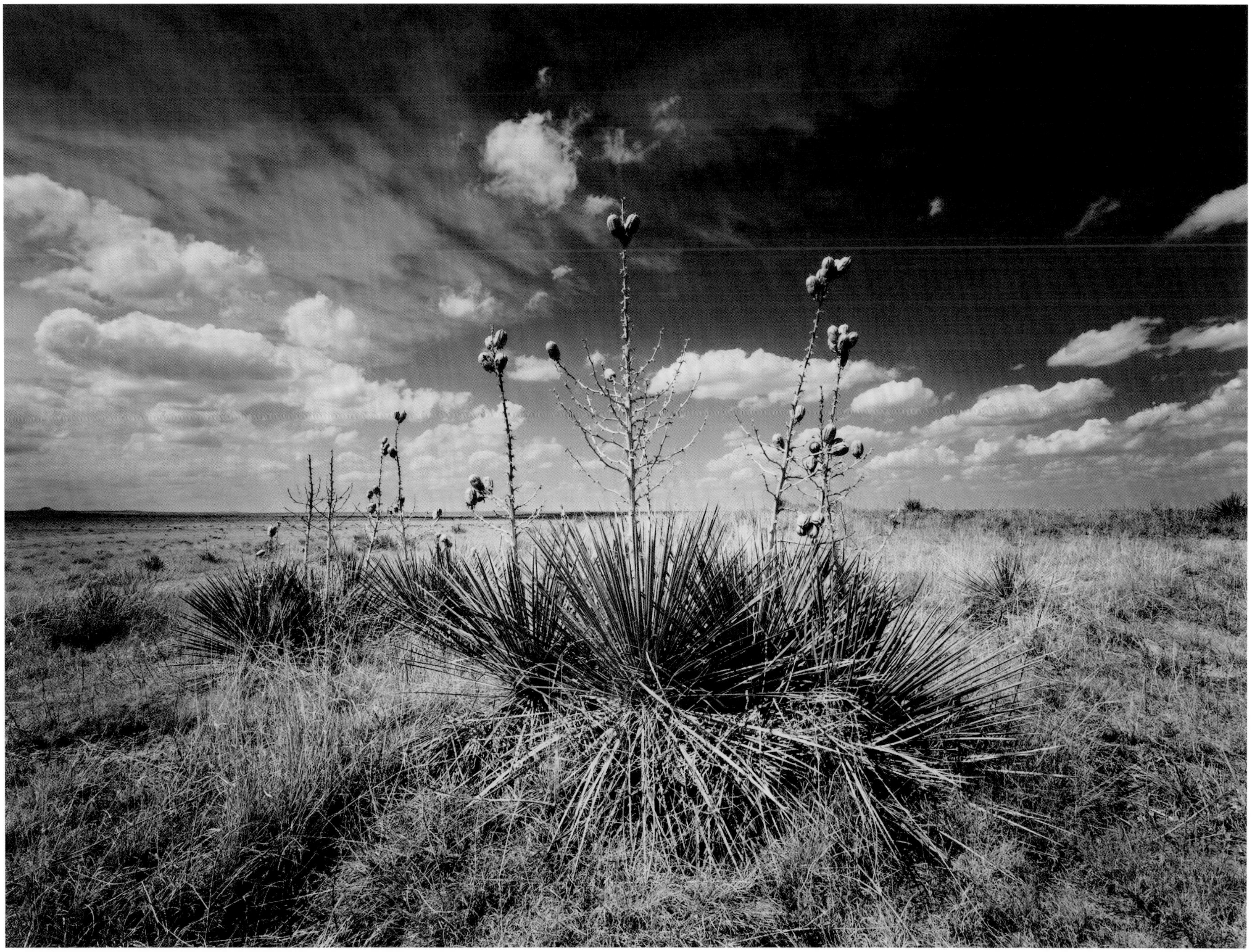

Fence, Kiowa National Grasslands, New Mexico, 2005

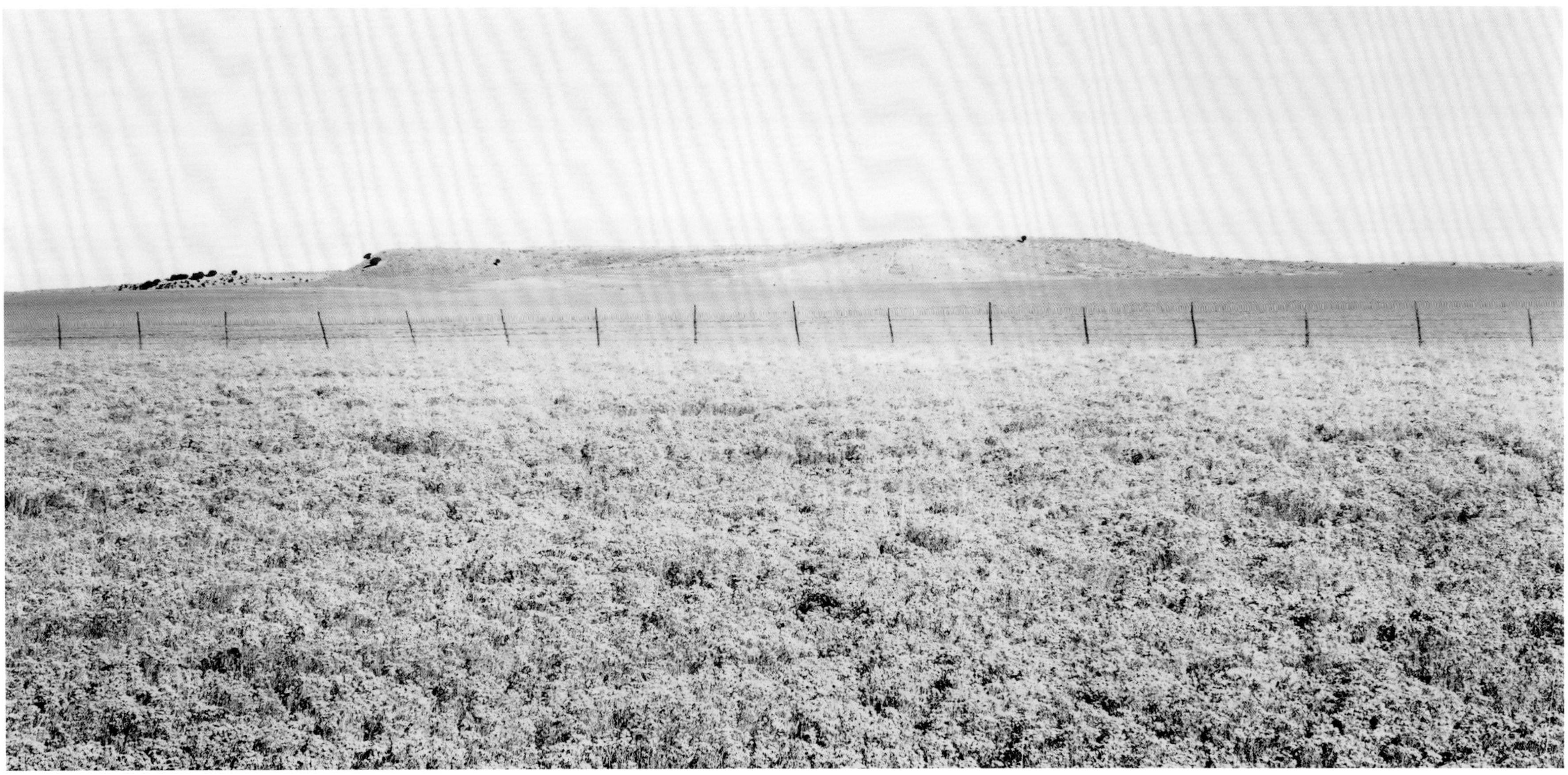

Windmill, Pawnee National Grasslands, Colorado, 2006

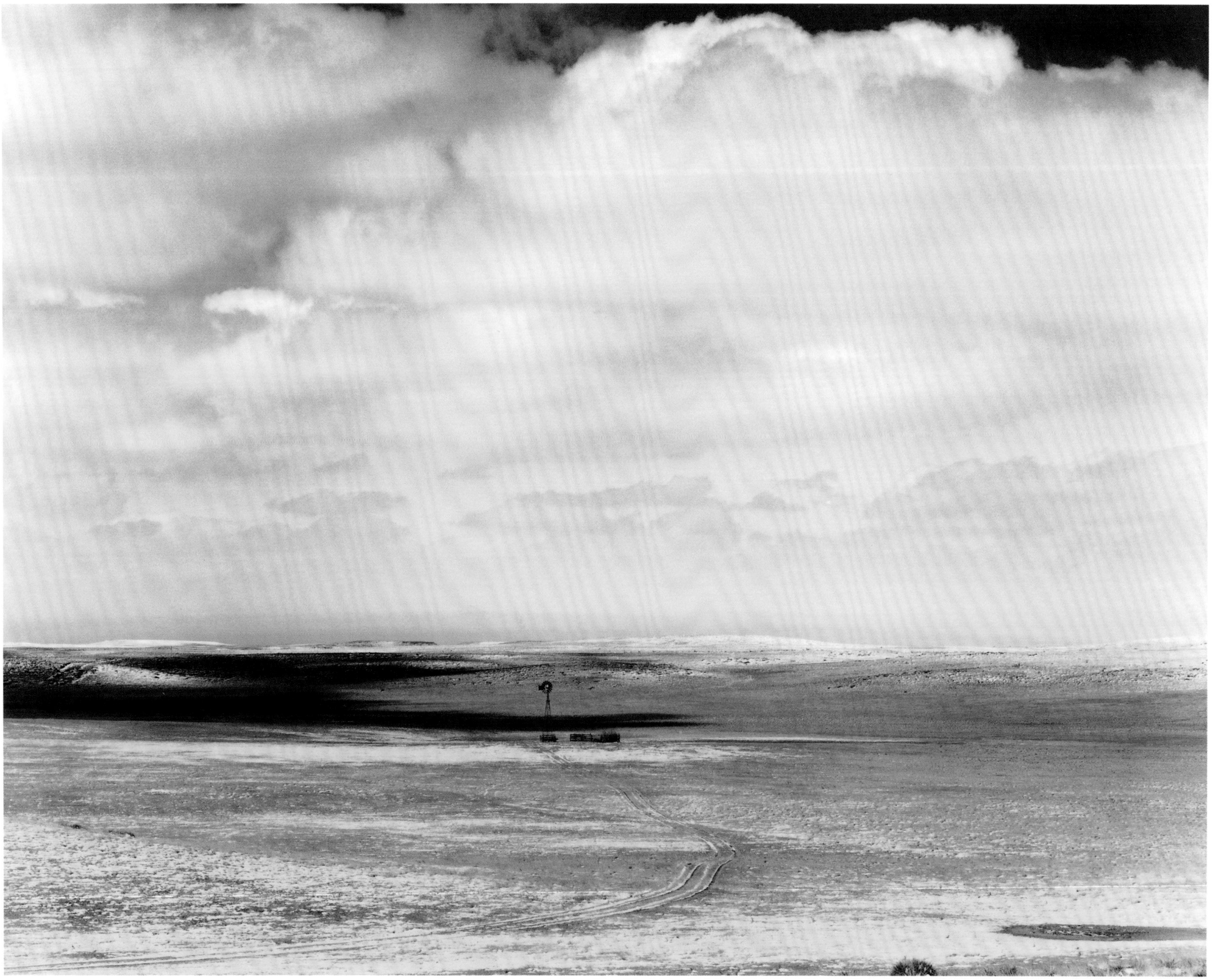

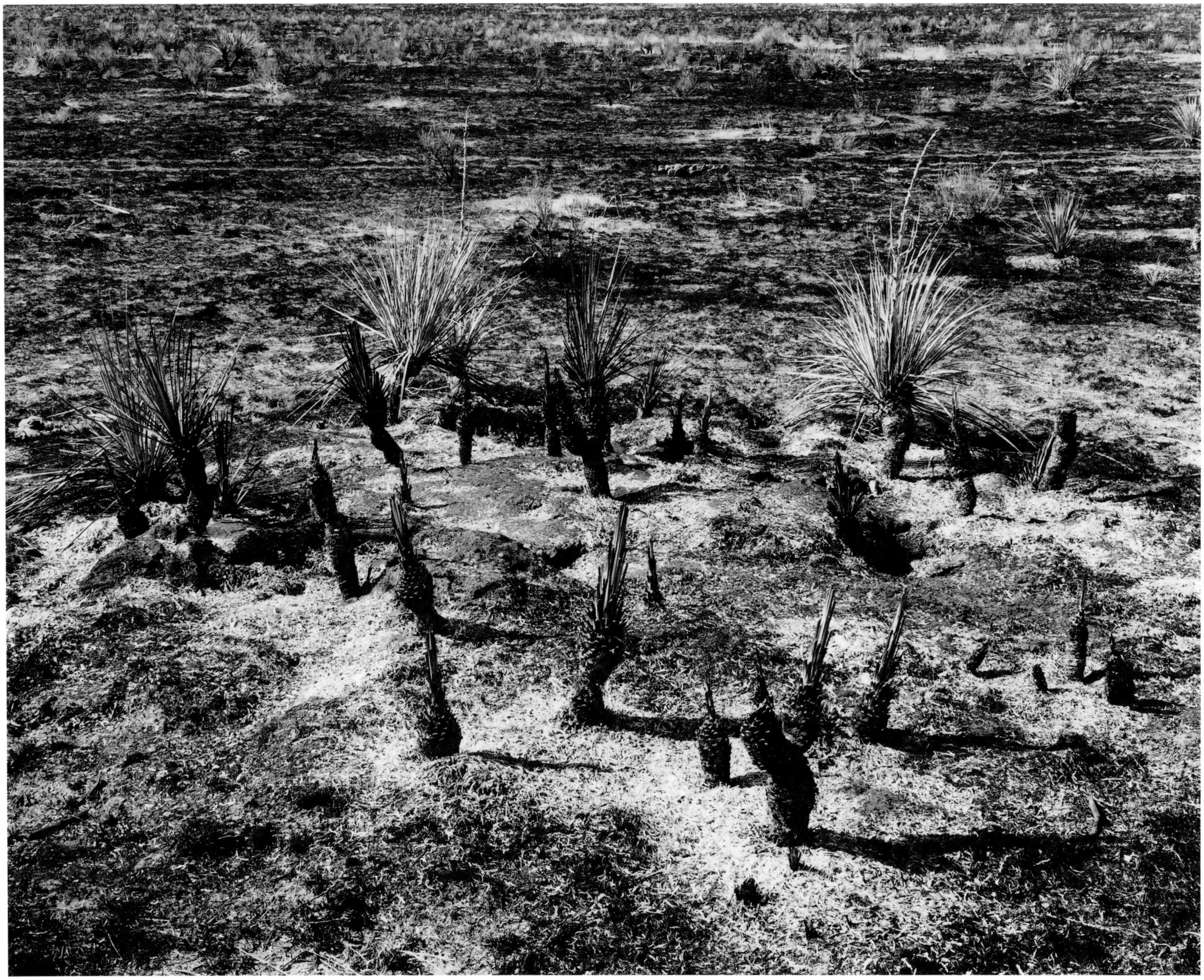

Burn Area, Comanche National Grasslands, Colorado, 2007

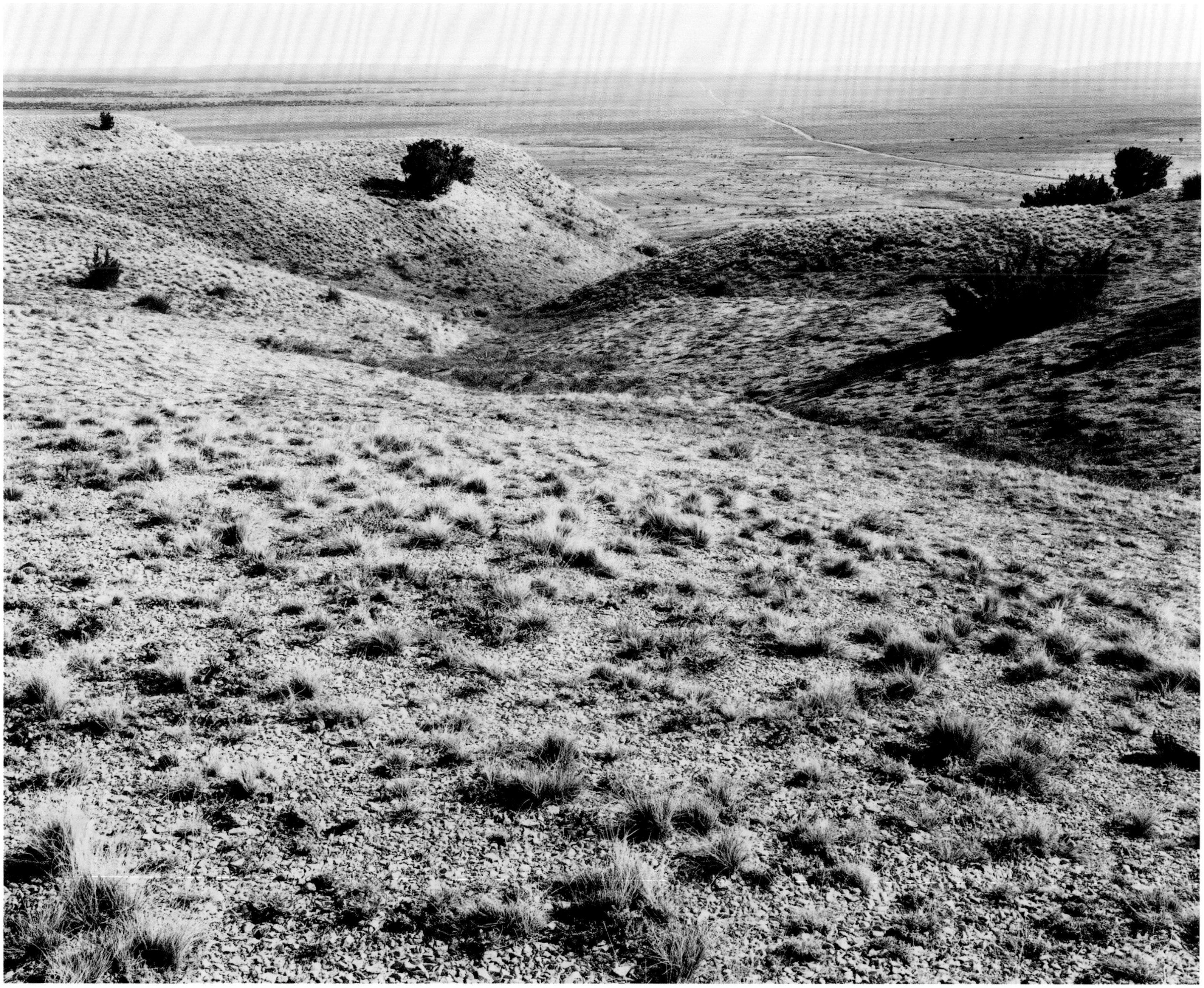

Shale Hills, Comanche National Grasslands, Colorado, 2007

Comanche National Grasslands, Colorado, 2006

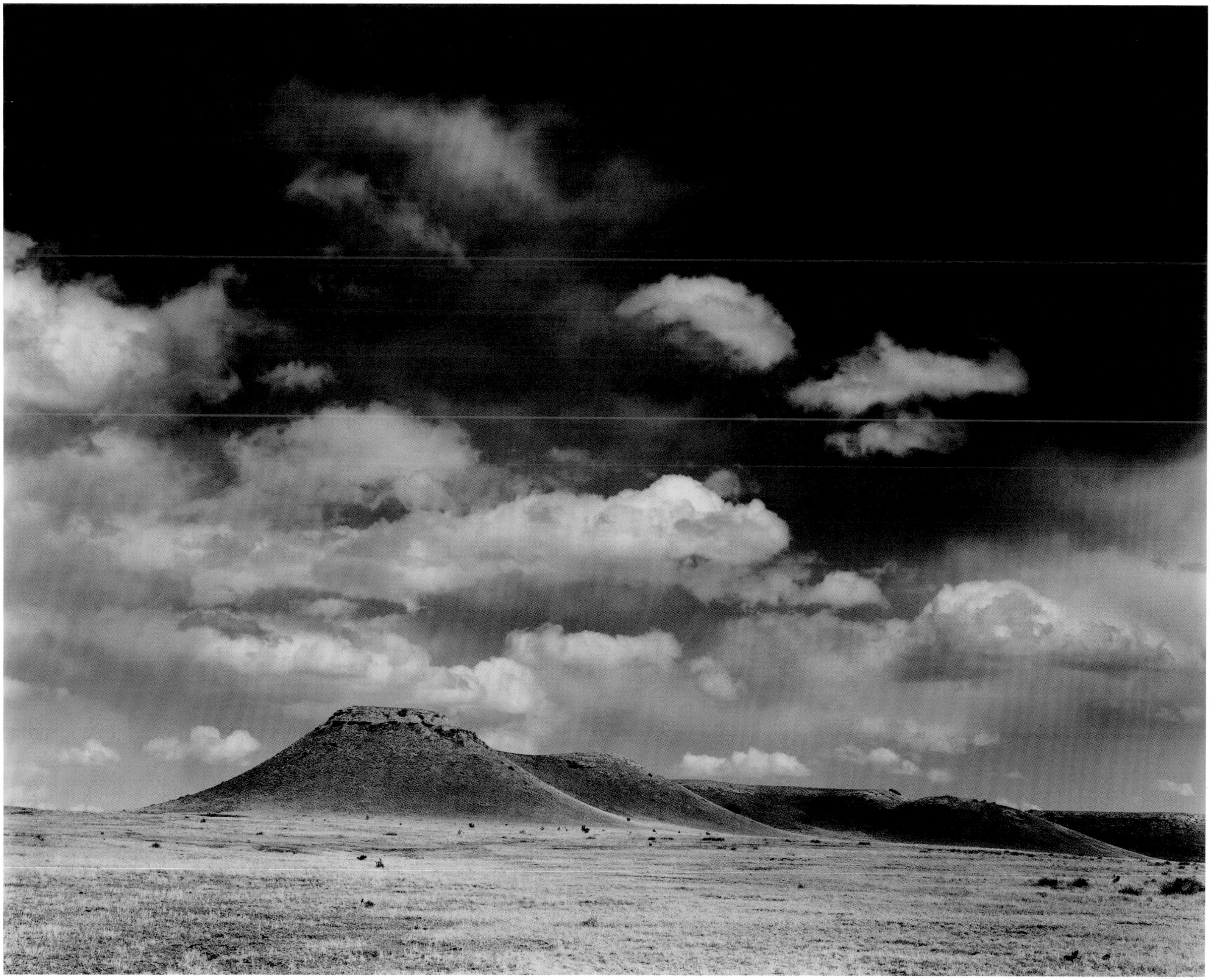

Moonrise, Thunder Basin National Grasslands, Wyoming, 2008

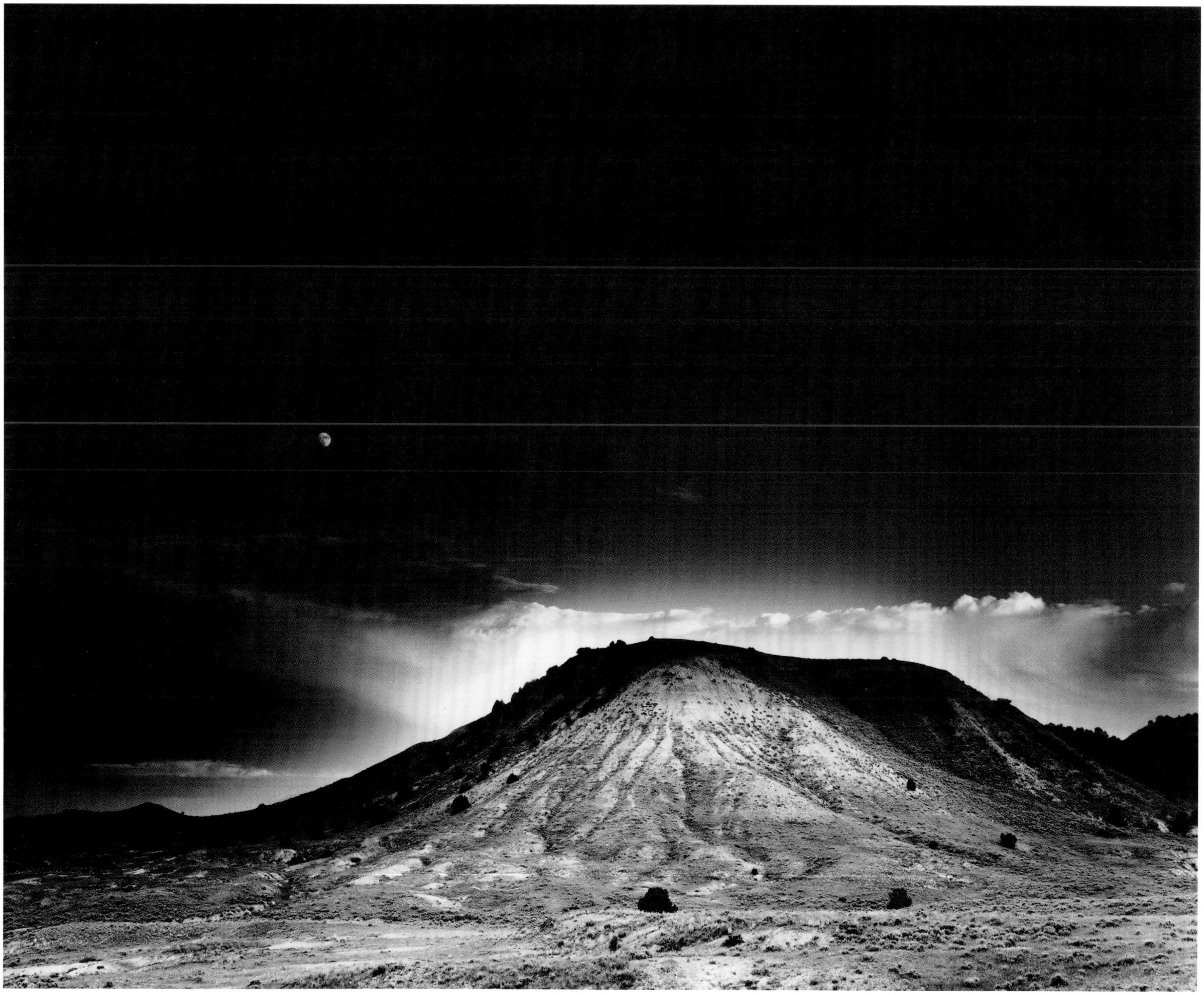

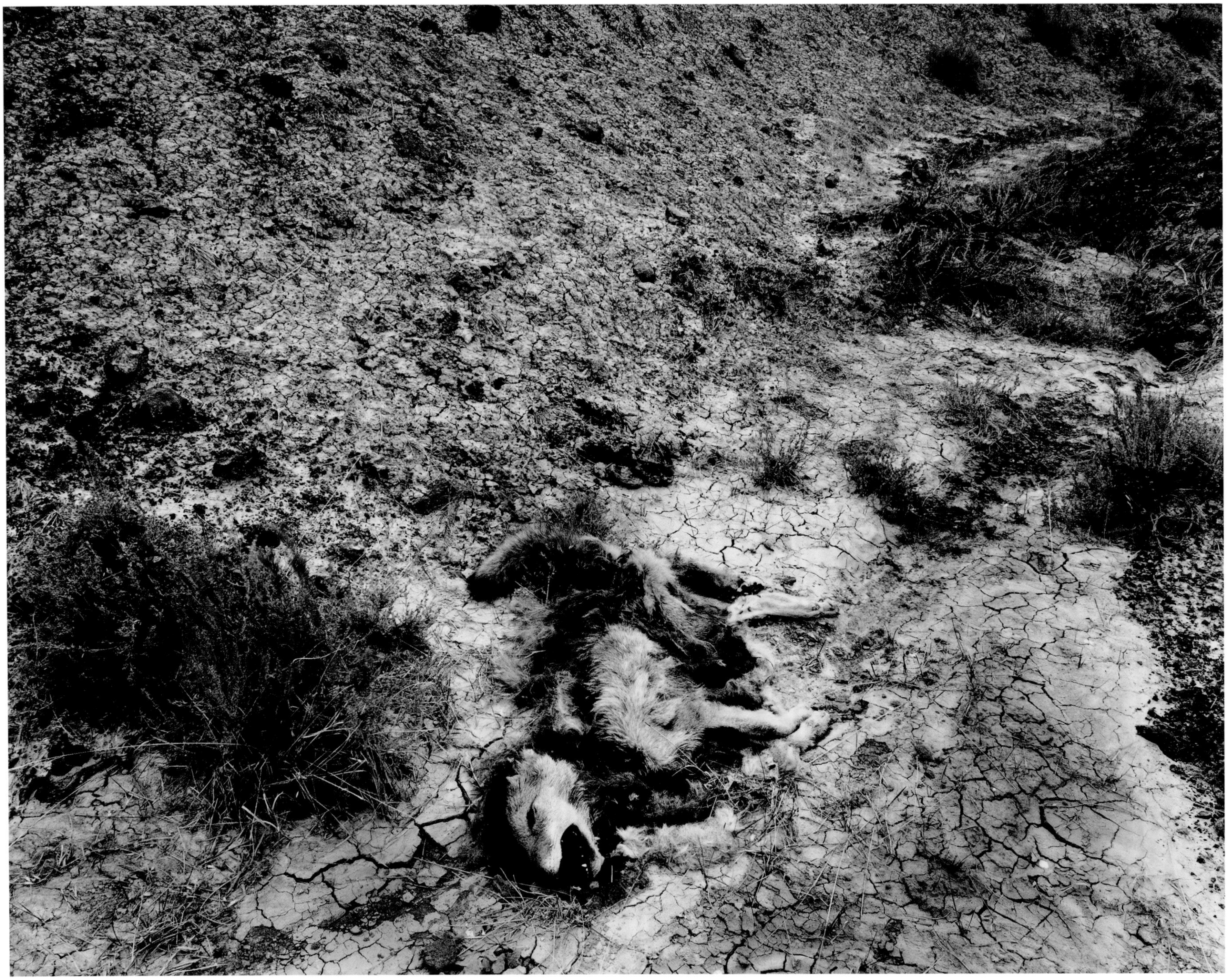

Coyote, Thunder Basin National Grasslands, Wyoming, 2006

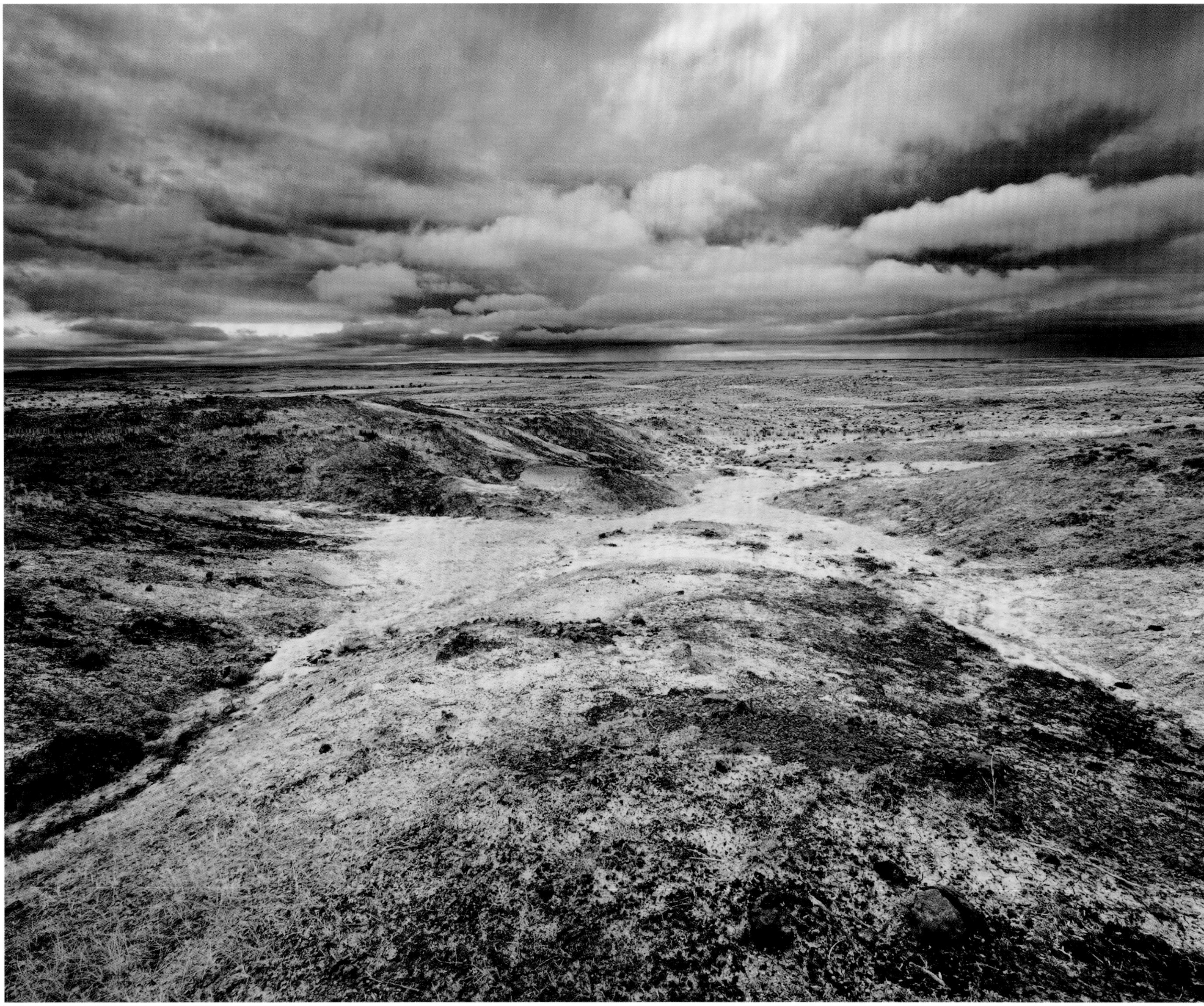

Thunder Basin National Grasslands, Wyoming, 2006

Thunder Basin National Grasslands, Wyoming, 2006

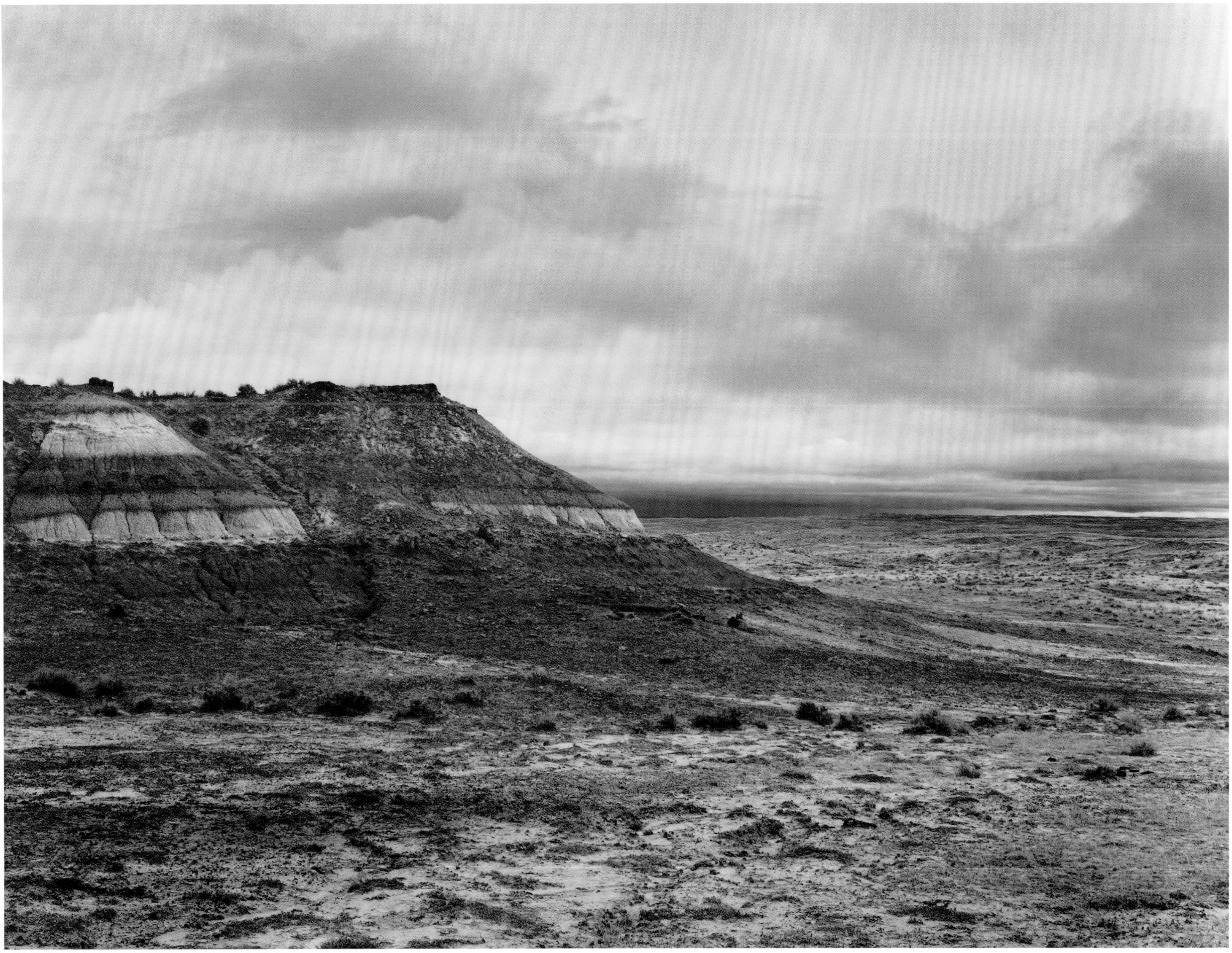

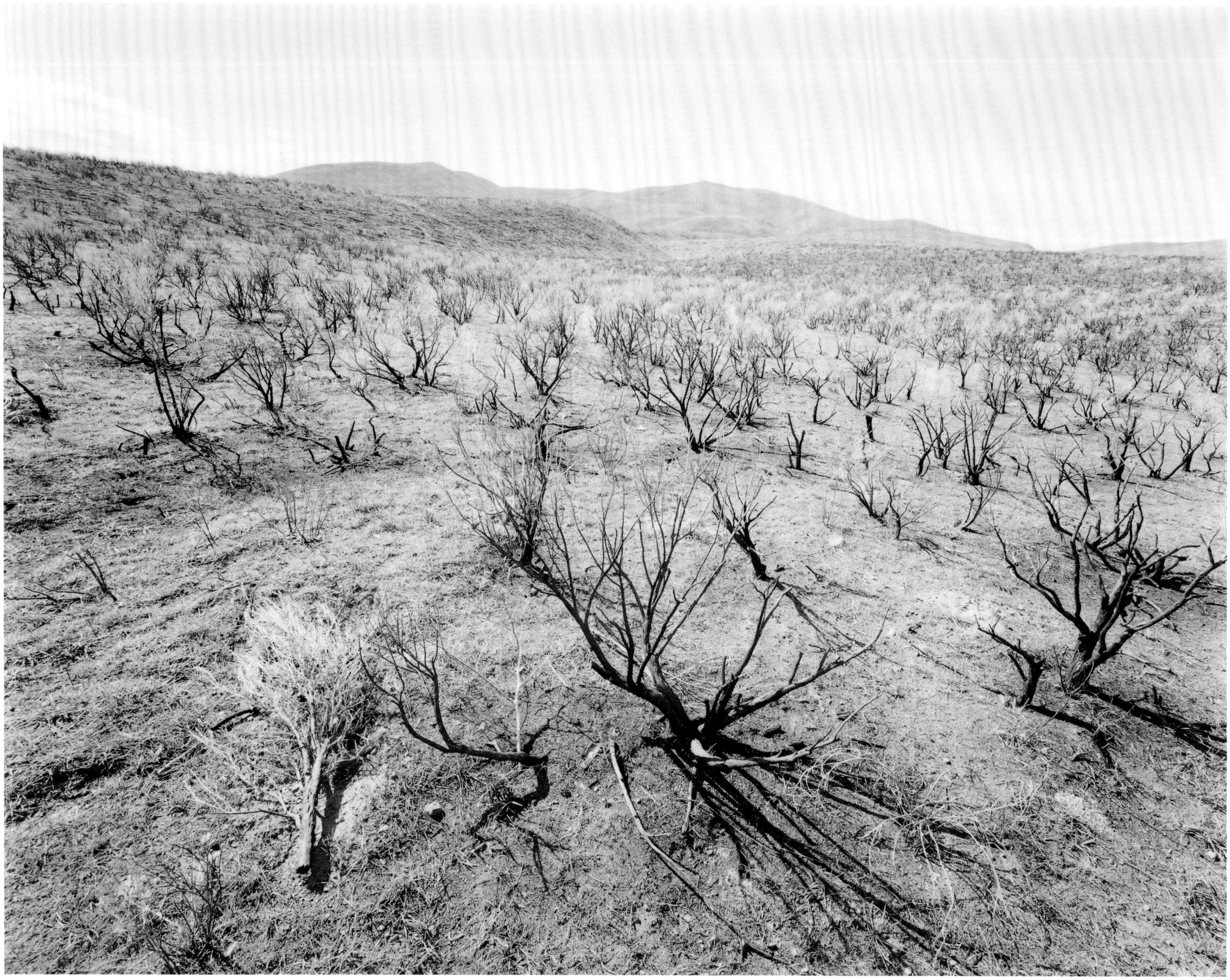

Burn Area, Curlew National Grasslands, Idaho, 2006

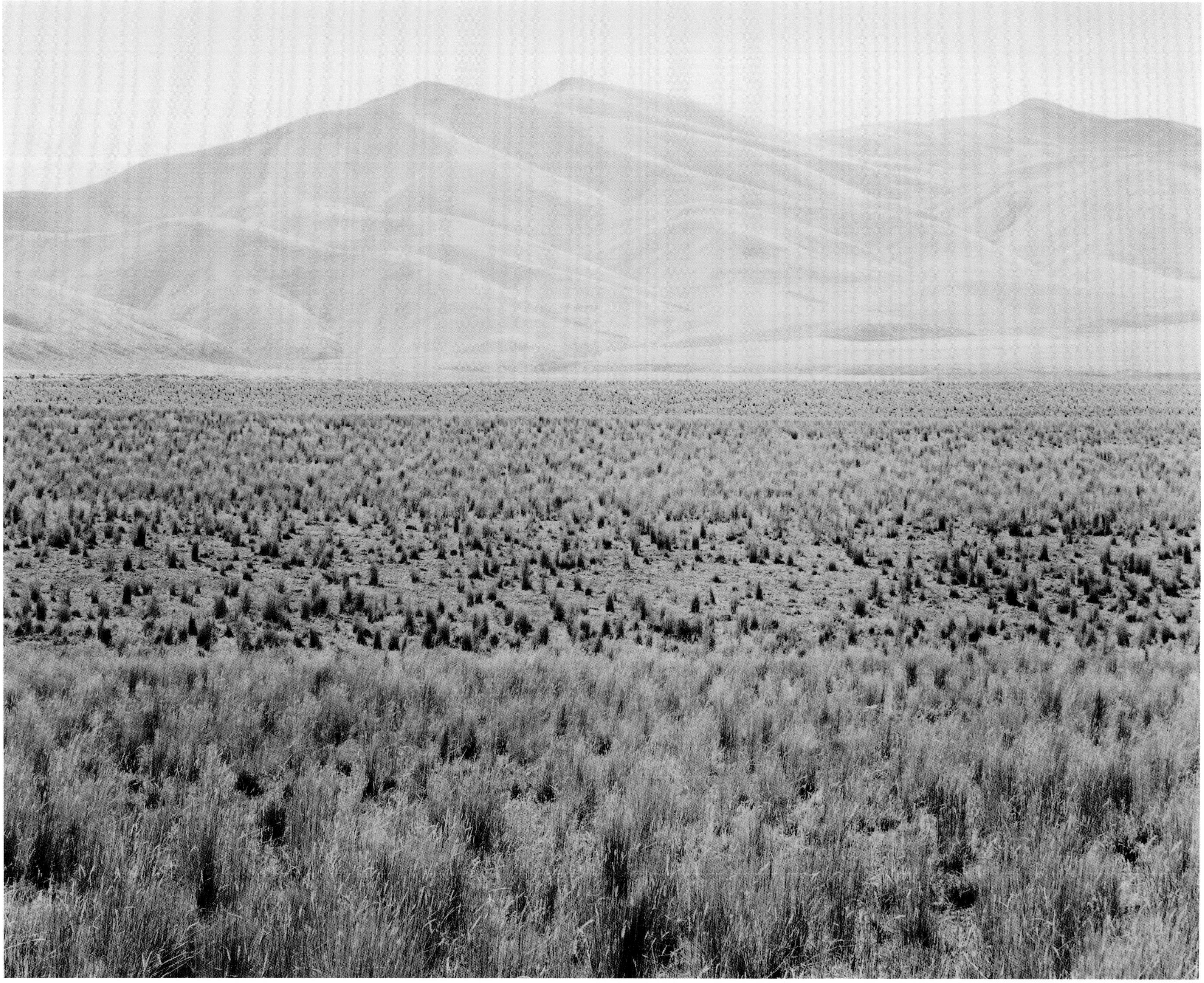

Bunchgrass, Curlew National Grasslands, Idaho, 2006

Curlew National Grasslands, Idaho, 2006

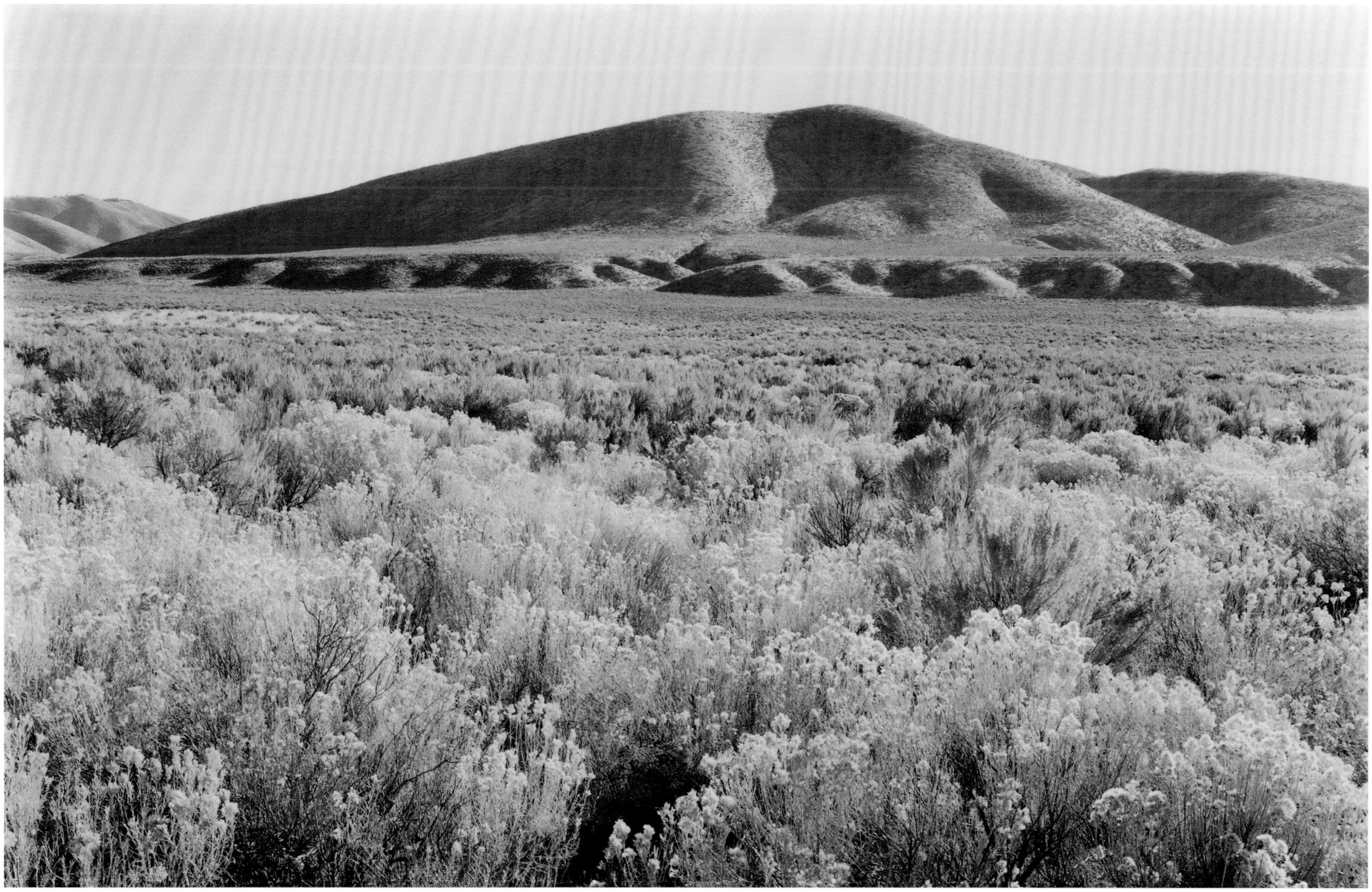

Lone Tree, Zumwalt Prairie, Oregon, 2007

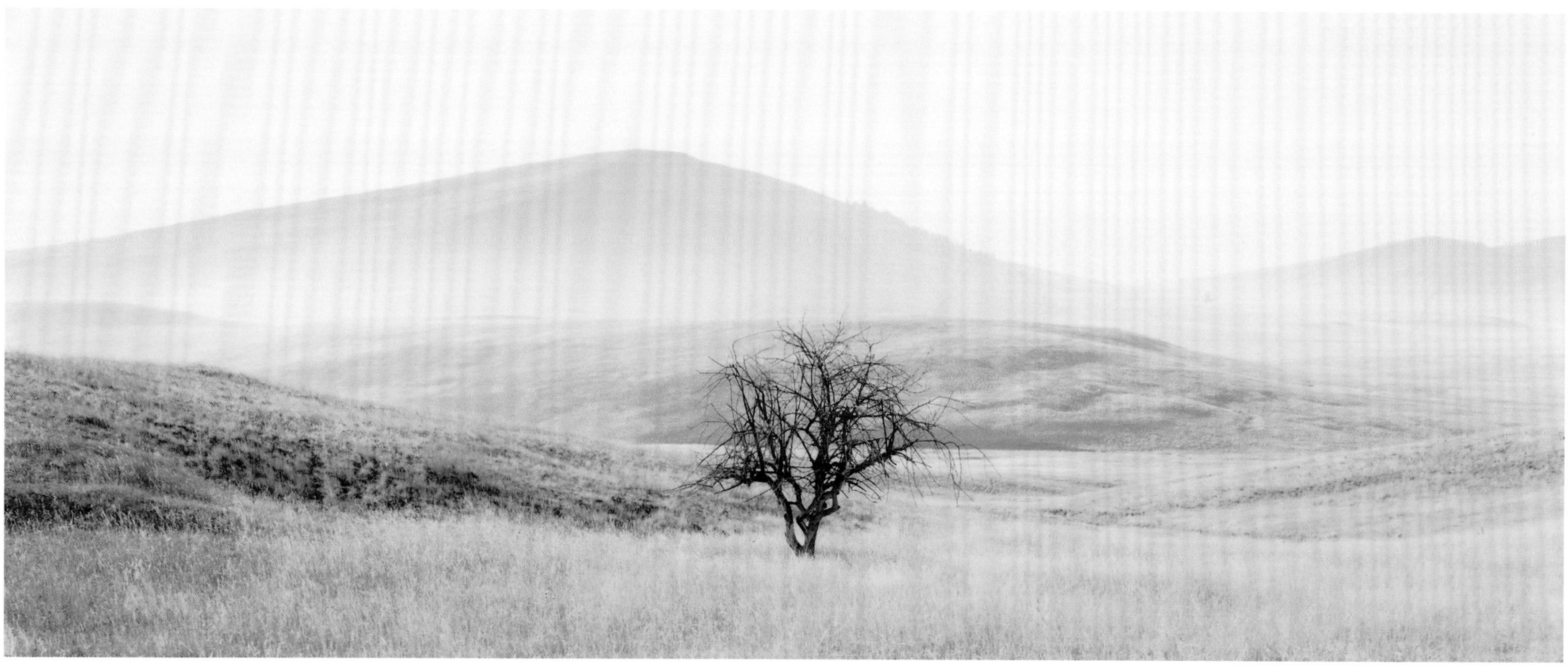

Tree and Grass, Crooked River National Grasslands, Oregon, 2006

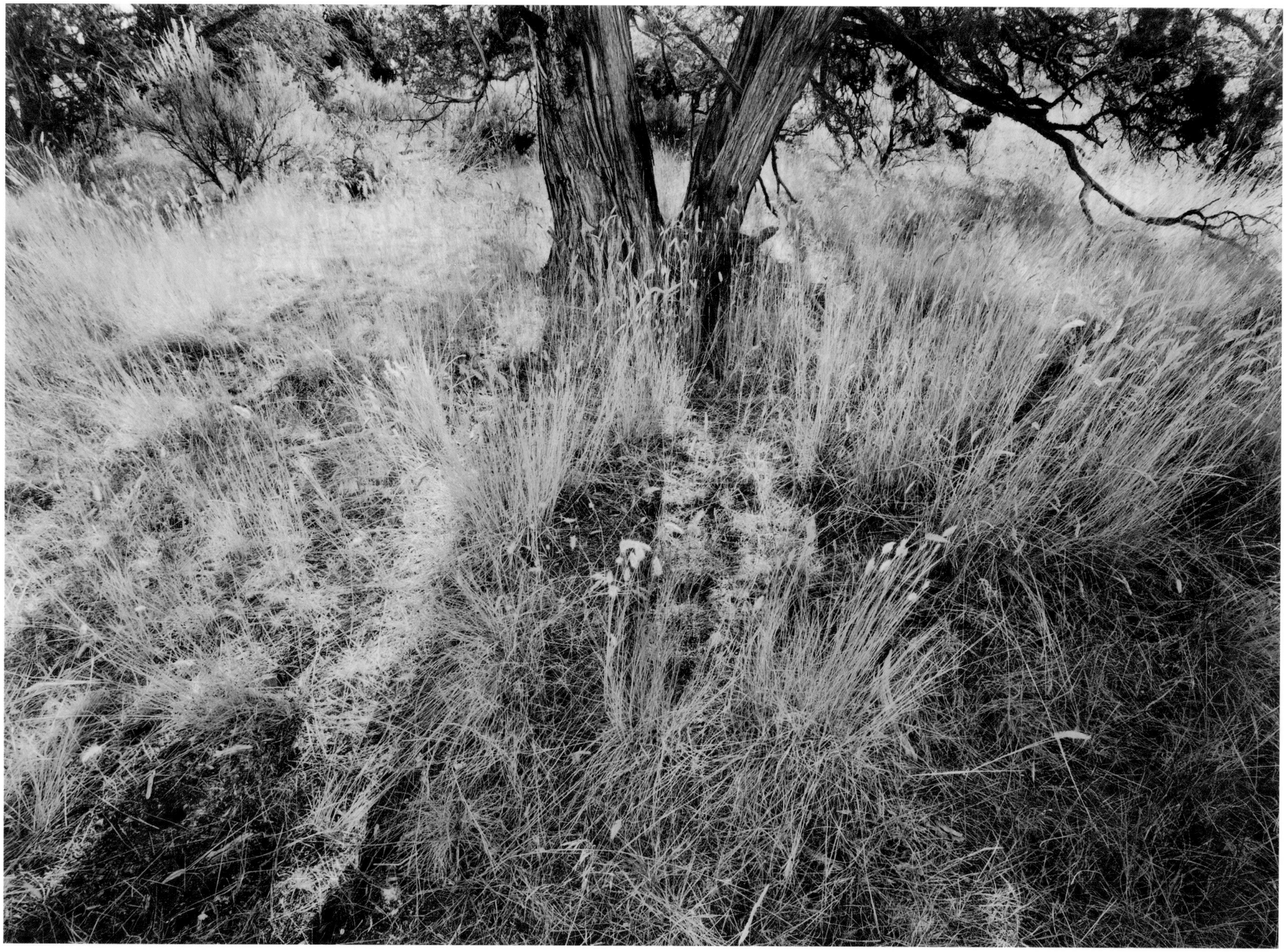

Sunrise, Crooked River National Grasslands, Oregon, 2006

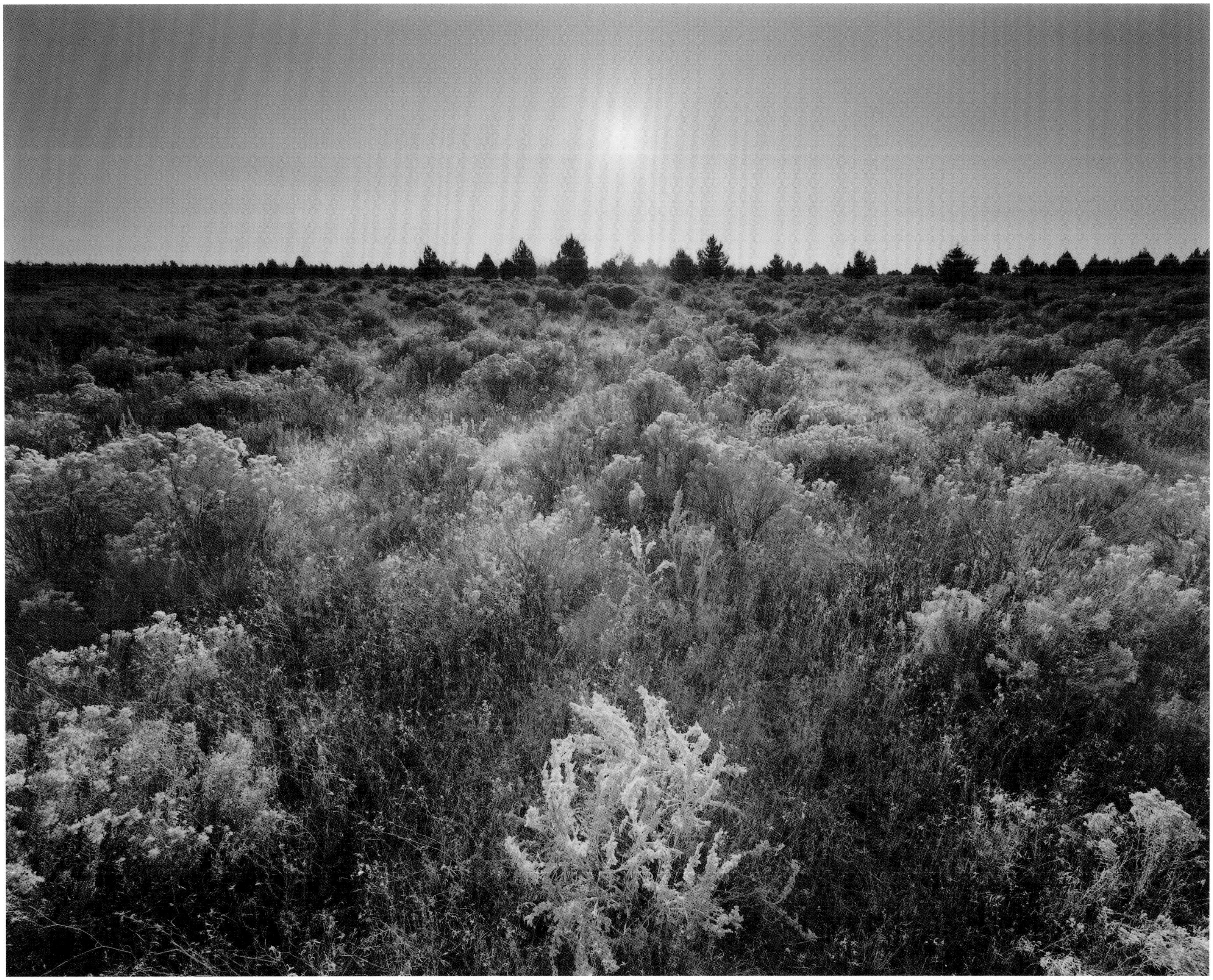

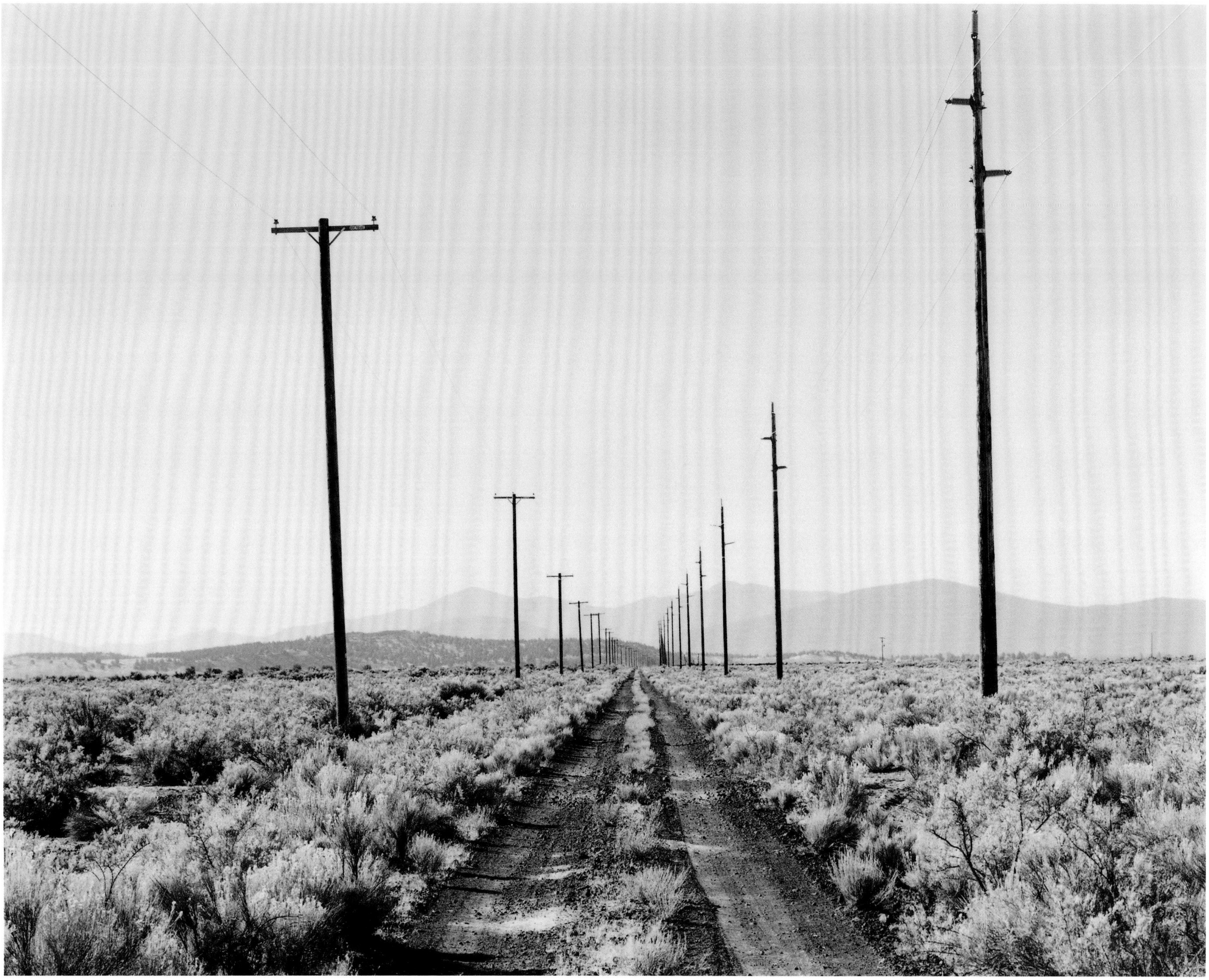

Power Lines, Butte Valley National Grasslands, California, 2008

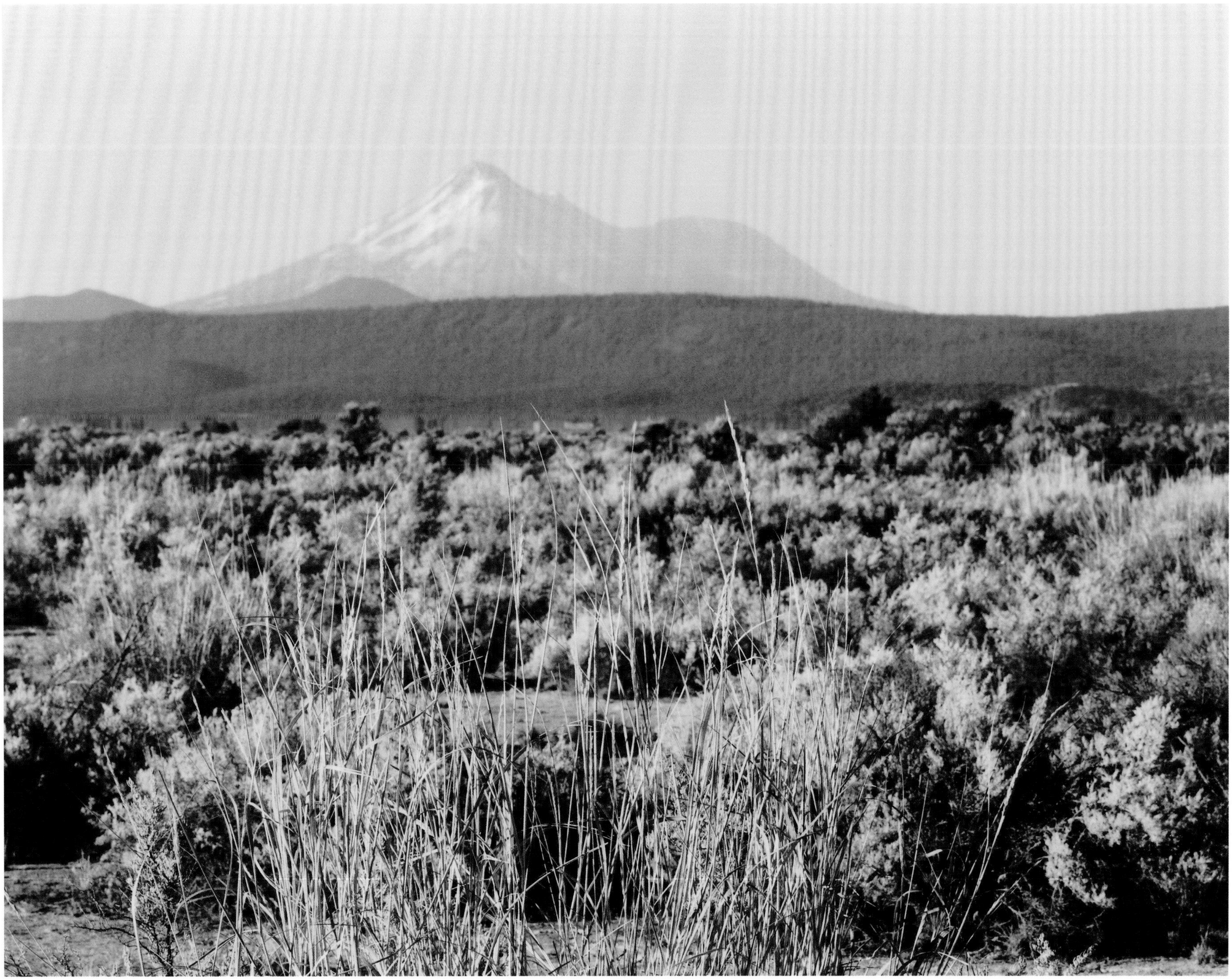

Butte Valley National Grasslands, California, 2008

Butte Valley National Grasslands, California, 2008

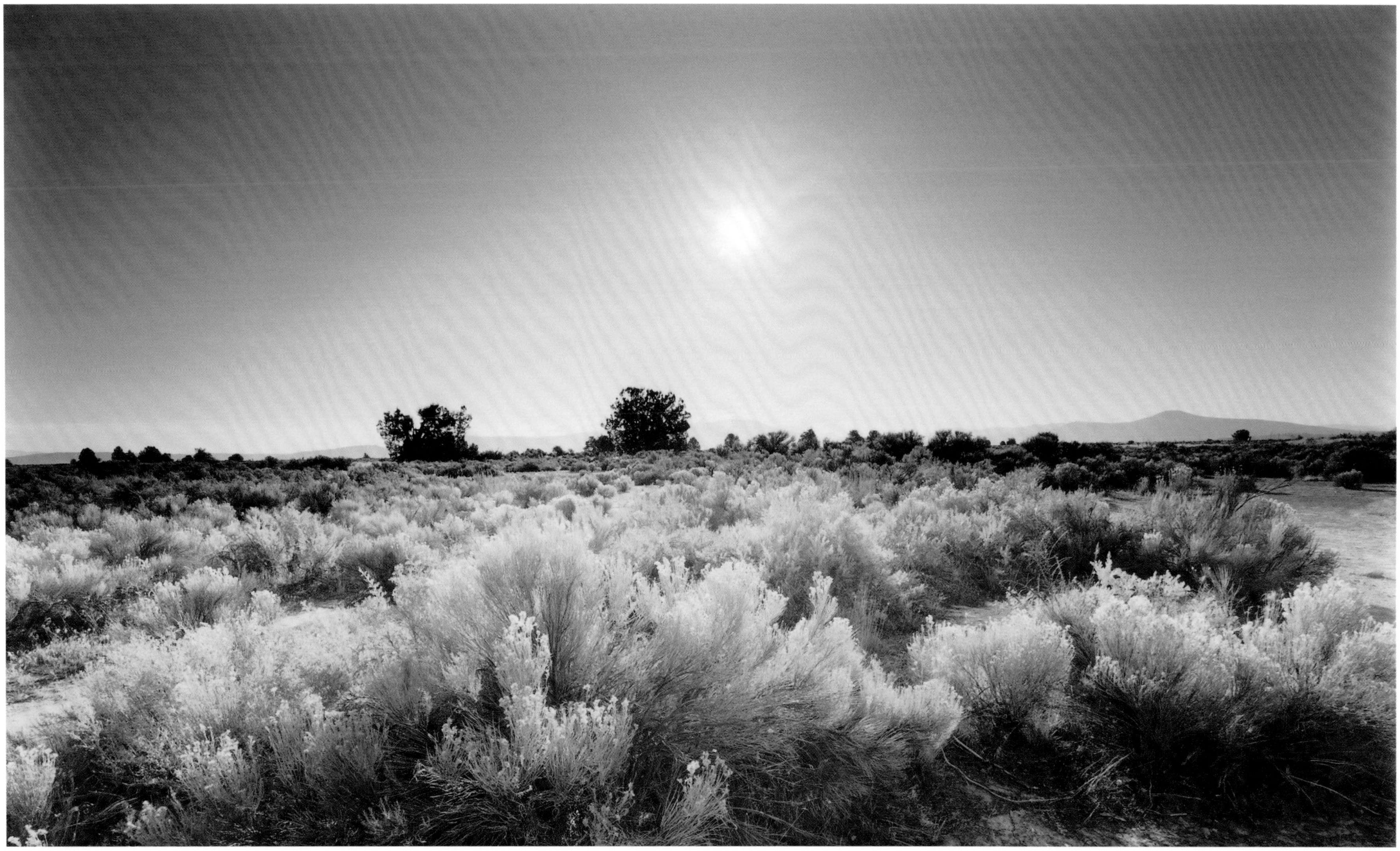

Grasslandus Interruptus

This book had to be called *Grasslands* because of discontinuity. The truthful S at the end of the word reinforces our knowledge that a once-vast *Grassland* existed on this continent before the onset of Manifest Destiny. It reminds us that we decimated the bison, rammed railroads through, overran rangeland with bovines from a wetter Europe, imposed the cadastral grid, turned sod upside down, and continue to till and spray to the far horizons. Great moments in fragmentation!

In June 2007 I visited my old workshop-teaching pard Kirby at his new digs near Santa Fe. On the heels of his Anasazi and Wheat Country achievements, I asked him, "What's next?" "National Grasslands—all of them," was his reply as he brought out several new silver prints. "Damn, Don," I said, "I hope you find a tree out there."

Well, he found the tree—in fact, dozens of them. Plus swales, draws, ridges, buttes, clouds, shadows, windstorms, firestorms, rainstorms. Most of all, the beautiful, life-sustaining grasses. I had hoped to see his variation on the Albrecht Dürer 1503 painting known as The Large Turf. "Couldn't do it. The tall grass was never still enough."

I love to be horseback in the intermountain grasslands around here. Like the ranch that I took Don to back in 1998, where the cattle are located and moved in ways that benefit the range, and the expansive stand of tall native Sacaton grass is burned regularly. We know now that Indians, using hunting pressure and fire, once "managed" the enormous bison herds of the Great Plains. The score-or-so of National Grasslands scattered about the West are being managed again. Managed and monitored. Among the restoration approaches and strategies: fire, cows, bison.

Don Kirby has found for us an archipelago of hope in an overstressed land.

Jay Dusard
Cochise County, Arizona, 2009

Notes to the Plates

GRASSLANDS

2871 NE Alamada Street
Portland, Oregon 97212

Printed in China
ISBN 978-1-59005-264-8